Microplastics and Nanoplastics in Aquatic Environments

Edited by Monique Mancuso and Bilal Mghili

Published in London, United Kingdom

Microplastics and Nanoplastics in Aquatic Environments
http://dx.doi.org/10.5772/intechopen.1009071
Edited by Monique Mancuso and Bilal Mghili

Contributors
Chengsong Ye, Divya Tyagi, Egara Wilfred Oviemuno, Etaware Peter Mudiaga, Jayendra Kumar Singh, John Dennis, Kassim Chabi, Mahmoud Gad, Mrinmoy Mandal, Naveen Kumar, Neveen M. Rizk, Nisha Yadav, Oviemuno Bridget Gwa, Rajendran Rajaram, Sourav Bhattacharya, Tapan Kumar Mistri, Xin Yu

First published in London, United Kingdom, 2026 by IntechOpen
IntechOpen is the global imprint of INTECHOPEN LIMITED, registered in England and Wales, registration number: 11086078, 167-169 Great Portland Street, London, W1W 5PF, United Kingdom

For EU product safety concerns: IN TECH d.o.o., Prolaz Marije Krucifikse Kozulić 3, 51000 Rijeka, Croatia, info@intechopen.com or visit our website at intechopen.com.

British Library Cataloguing-in-Publication Data
A catalogue record for this book is available from the British Library

Microplastics and Nanoplastics in Aquatic Environments
Edited by Monique Mancuso and Bilal Mghili
p. cm.

This title is part of the Environmental Sciences Book Series, Volume 47
Topic: Pollution
Series Editor: J. Kevin Summers
Topic Editors: Ismail Rahman and Zinnat Ara Begum

Print ISBN 978-1-83634-587-9
Online ISBN 978-1-83634-586-2
eBook (PDF) ISBN 978-1-83634-588-6
ISSN 2754-6713

If disposing of this product, please recycle the paper responsibly.

IntechOpen Book Series

Environmental Sciences

Volume 47

Aims and Scope of the Series

Scientists have long researched to understand the environment and man's place in it. The search for this knowledge grows in importance as rapid increases in population and economic development intensify humans' stresses on ecosystems. Fortunately, rapid increases in multiple scientific areas are advancing our understanding of environmental sciences. Breakthroughs in computing, molecular biology, ecology, and sustainability science are enhancing our ability to utilize environmental sciences to address real-world problems.

The four topics of this book series - Pollution; Environmental Resilience and Management; Ecosystems and Biodiversity; and Water Science - will address important areas of advancement in the environmental sciences. They will represent an excellent initial grouping of published works on these critical topics.

Meet the Series Editor

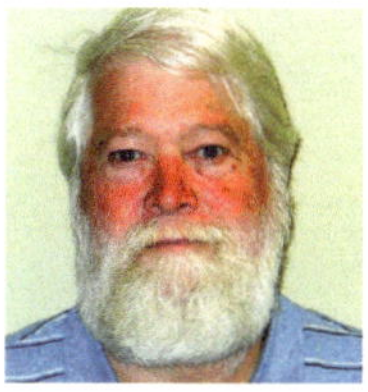

J. Kevin Summers has been working as a Senior Research Ecologist at the Environmental Protection Agency's (EPA) Gulf Ecosystem Measurement and Modeling Division. He worked with colleagues in the Sustainable and Healthy Communities Program to develop an index of community resilience to natural hazards, an index of human well-being that can be linked to changes in the ecosystem, social and economic services, and a community sustainability tool for communities with populations under 40,000. He led research efforts for indicator and indices development. Dr. Summers is a systems ecologist and began his career at the EPA in 1989 and has worked in various programs and capacities. This includes leading the National Coastal Assessment in collaboration with the Office of Water, which culminated in the award-winning National Coastal Condition Report series (four volumes between 2001 and 2012), and which integrates water quality, sediment quality, habitat, and biological data to assess the ecosystem condition of the United States estuaries. He was acting National Program Director for Ecology for the EPA between 2004 and 2006. He has authored approximately 150 peer-reviewed journal articles, book chapters, and reports, and has received many awards for technical accomplishments from the EPA and from outside of the agency. Dr. Summers holds a BA in Zoology and Psychology, an MA in Ecology, and Ph.D. in Systems Ecology/Biology.

Meet the Volume Editors

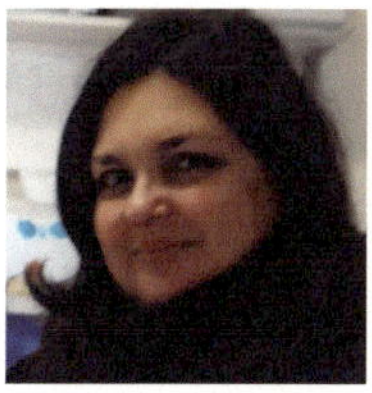

Monique Mancuso is a biologist and researcher at the Institute for Biological Resources and Marine Biotechnologies (IRBIM), CNR, where she has been working since 2001. Her research focuses on monitoring marine litter and assessing the ecological impacts of microplastics and nanoplastics on marine biodiversity, including fish, elasmobranchs, and invertebrates. She is dedicated to advancing knowledge on marine pollution and biodiversity conservation. She has coordinated and continues to coordinate several research projects, collaborating internationally with institutions such as the Polish Academy of Sciences and the University of Montenegro, and serves as Principal Investigator for a sub-task within the National Recovery and Resilience Plan (PNRR). She is also affiliated with the National Biodiversity Future Center (NBFC). Throughout her career, she has participated in 30 research projects and five oceanographic expeditions in the Mediterranean Sea. She has 71 Scopus-indexed publications with an h-index of 24 and is active as an editor and reviewer for numerous scientific journals.

Dr. Bilal Mghili is a marine biologist specializing in marine ecosystems and conservation. He completed his Ph.D. in Marine Biology at the University of Abdelmalek Essaâdi, focusing on jellyfish blooms and their impacts in the Moroccan Mediterranean. Dr. Mghili has published numerous peer-reviewed articles in leading journals and has contributed to international efforts to protect the marine environment. In addition to his research, he serves as a peer reviewer for several leading marine science journals, contributing his expertise to advancing the field. He currently works as a senior researcher at the University of Abdelmalek Essaâdi, where he investigates the effects of plastic pollution on the marine environment.

Contents

Preface

This book, *Microplastics* and *Nanoplastics in Aquatic Environments,* aims to provide a comprehensive and multidisciplinary overview of the growing threat posed by microplastics and nanoplastics in aquatic ecosystems. This global-scale problem has been attracting increasing attention from the scientific community, regulatory authorities, and the general public. The spread of these particles in the aquatic environments has now become ubiquitous, and their effects, both ecological and on the health of organisms, are complex and often underestimated.

The book examines the ecotoxicological effects of microplastics and nanoplastics, highlighting how they can interfere with the fundamental biological processes of aquatic organisms. Due to their extremely small size and persistence in the environment, these particles can penetrate organs and tissues, causing oxidative stress, inflammation, and functional alterations. Particular attention is given to interactions between micro- and nanoplastics and other chemical or biological contaminants in ecosystems, which can amplify overall toxicity and pose a significant threat to biodiversity and the stability of aquatic ecosystems.

A key chapter of the book focuses on the microscopic perspective, illustrating how nanoplastics, due to their size of less than 100 nanometres, can cross cellular membranes and biological barriers, acting as genuine "intrusive agents" at the molecular level. This discussion allows an understanding not only of the direct damage to individual organisms but also of systemic consequences along the food chain, extending to potential effects on higher organisms and humans, particularly through the consumption of contaminated aquatic species.

Alongside the biological and chemical approach, the book includes an ethical and philosophical reflection on the human role in the microplastic crisis. It examines collective responsibility towards aquatic ecosystems, raising fundamental questions about environmental justice and the consumption and production choices that have led to the current state of emergency. Readers are encouraged to consider not only the scientific and health implications but also the moral consequences of human actions on shared habitats and future generations.

The book is therefore presented as a valuable resource for researchers, students, and professionals in environmental science and public health, providing conceptual tools and up-to-date data to understand the complexity of the problem and develop mitigation and management strategies. The interdisciplinary insights in the volume underscore the urgency of addressing microplastics and nanoplastics through innovative approaches that integrate ecology, toxicology, chemistry, environmental policy, and ethics.

A sincere note of thanks goes to Maja Božičević for her invaluable support over these months, which made the completion of this work possible.

In conclusion, *Microplastics and Nanoplastics in Aquatic Environments* represents a significant contribution to the scientific literature, combining analytical rigour, experimental insight, and critical reflection, and provides a comprehensive overview of the environmental threat posed by plastic particles. Its goal is to raise awareness, inform, and stimulate concrete actions to protect the health of aquatic ecosystems and, by extension, the communities that depend on them.

Monique Mancuso
Department of Earth System Sciences and
Environmental Technologies (DISTTA),
Institute for Research on Biological Resources and
Marine Biotechnology (IRBIM),
National Research Council (CNR),
Messina, Italy

Bilal Mghili
Faculty of Sciences,
Department of Biology at Abdelmalek Essaadi,
Laboratory for Environmental Sciences and
Coastal Biology (LESCB, URL-CNRST No. 18),
Abdelmalek Essaâdi University,
Tétouan, Morocco

Chapter 1

Ecotoxicological and Health Impacts of Micro- and Nanoplastics

Sourav Bhattacharya and Rajendran Rajaram

Abstract

Microplastics (MPs; <5 mm) and nanoplastics (NPs; <1 μm) have emerged as pervasive and persistent contaminants in aquatic ecosystems, posing significant risks to both environmental and human health. This chapter critically explores the ecotoxicological mechanisms and biological consequences of MPs/NPs across multiple levels of biological organization. Key factors such as particle size, polymer type, surface properties, and environmental weathering play crucial roles in determining their bioavailability, transport, and toxicity. MPs and NPs have been shown to induce oxidative stress, inflammation, genotoxicity, endocrine disruption, and neurotoxicity in aquatic organisms, with detrimental effects on reproduction, development, and behavior. These effects cascade through trophic levels, potentially altering ecosystem stability and function. A major emerging concern is the formation of the plastisphere, a dynamic microbial community on plastic surfaces that serves as a reservoir for pathogens and antibiotic resistance genes. This plastisphere contributes to the horizontal gene transfer of resistance traits and may facilitate disease emergence, thus bridging ecological and public health risks. Humans are increasingly exposed to MPs/NPs through seafood consumption, drinking water, inhalation, and occupational contact. While epidemiological data remain limited, *in vivo* and *in vitro* studies suggest possible links to inflammation, immune dysregulation, reproductive toxicity, and gut microbiome disturbances. This chapter also identifies critical knowledge gaps in current ecotoxicological and human health assessments, emphasizing the need for standardized methodologies, interdisciplinary collaboration, and regulatory integration. Advancing our understanding of MPs/NPs is essential for mitigating risks, guiding sustainable plastic use, and informing international environmental and public health policies.

Keywords: microplastics, nanoplastics, plastisphere, bioaccumulation, environmental contaminants

1. Introduction

Over the past few decades, the pervasive proliferation of microplastics (MPs; <5 mm) and nanoplastics (NPs; <1 μm) in aquatic ecosystems has evolved from a nascent environmental issue into a pressing global crisis, affecting both ecological integrity and public health [1–3]. These particulate pollutants, originating from primary sources such as personal care products, synthetic fibers, and industrial abrasives, as well as from the fragmentation of larger macroplastic debris, are characterized by their physicochemical persistence, environmental ubiquity, and high bioactivity [1, 3]. Their diminutive size and elevated surface area-to-volume ratio confer considerable reactivity and enable complex interactions with both biotic organisms and abiotic substrates across freshwater, estuarine, and marine systems [4, 5].

MPs and NPs exhibit the capacity to elicit adverse toxicological effects across hierarchical levels of biological organization, from subcellular and molecular domains to entire ecosystems [6]. Upon environmental entry, these particles undergo rapid colonization by diverse microbial assemblages and readily adsorb natural organic matter, heavy metals, persistent organic pollutants (POPs), and other co-contaminants [7, 8]. This process imparts a potent "vector effect," amplifying the toxicity of the composite particle-contaminant complexes and enabling the transboundary transport of hazardous substances across trophic networks [9, 10]. Empirical evidence now demonstrates that MPs and NPs can breach cellular membranes, traverse epithelial barriers, and accumulate in vital organs such as the gastrointestinal tract, liver, gills, and central nervous system in a variety of aquatic taxa [11, 12].

Laboratory experiments and field-based studies have consistently shown that exposure to MPs and NPs induces oxidative stress via excessive generation of reactive oxygen species (ROS), culminating in lipid peroxidation, genomic instability, mitochondrial dysfunction, and programmed cell death (apoptosis) [13–15]. Furthermore, these particles have been implicated in disrupting endocrine homeostasis, impairing reproductive output, inducing neurotoxicity, and altering behavior across multiple aquatic phyla, including zooplankton, mollusks, crustaceans, teleost fish, and amphibians [16, 17]. These physiological and behavioral impairments compromise organismal fitness and propagate through ecological networks, thereby undermining population stability, community structure, and ecosystem service delivery [17]. Unlike earlier reviews that primarily provided descriptive accounts, recent research has moved toward mechanistic insights. For instance, Ma et al. [11] demonstrated translocation of MPs/NPs in fish organs, while Bhat et al. [18] reported endocrine and reproductive disturbances in aquatic vertebrates. These findings reveal that particle size, shape, polymer type, and surface chemistry directly influence biological interactions.

The potential ramifications for human health are increasingly evident. Micro and nanoplastics have been detected in a wide range of exposure media, including seafood, bottled water, table salt, and even atmospheric deposition, indicating multiple routes of human uptake [19]. Following ingestion, inhalation, or dermal exposure, MPs and NPs may act as vectors for xenobiotics and microbial pathogens, with the potential to trigger inflammatory responses, genotoxic effects, endocrine perturbations, and dysbiosis of the gut microbiota [19, 20]. Of particular concern are high-risk populations, such as residents of coastal zones, children, and individuals employed in sectors with elevated plastic exposure, namely aquaculture, fisheries, wastewater treatment, and plastic recycling industries [20].

An emerging frontier in plastic-associated risk involves the microbial colonization of plastic surfaces, termed the plastisphere, which harbors complex biofilms comprising opportunistic pathogens (e.g., Vibrio spp.), antibiotic resistance genes (ARGs), and virulence determinants [21–23]. This microbial consortium raises significant ecological and epidemiological concerns due to its capacity to facilitate the environmental dissemination of infectious agents and contribute to the propagation of antimicrobial resistance (AMR) [24–26]. The introduction of plastisphere studies further complicates the risk scenario, as biofilms on MPs/NPs host opportunistic pathogens and antibiotic resistance genes, raising new public health concerns [20, 23].

This chapter delivers a critical, interdisciplinary synthesis of the ecotoxicological impacts and human health risks associated with micro and nanoplastics in aquatic systems. Drawing on the disciplines of environmental toxicology, analytical chemistry, microbial ecology, and epidemiology, we dissect the mechanistic underpinnings of particle toxicity, trophic transfer pathways, and bioaccumulation processes. Special focus is devoted to the emerging role of the plastisphere in mediating ecological interactions and disease dynamics. Ultimately, this work underscores the imperative for integrative research frameworks, evidence-driven policy interventions, and innovative mitigation strategies to confront the escalating threats posed by plastic contaminants in aquatic environments. Despite these advances, critical gaps remain, including insufficient data on long-term chronic exposure, lack of standard methodologies for nanoplastic detection, and weak integration of ecotoxicological findings with human epidemiological outcomes. Addressing these gaps requires an interdisciplinary One Health framework that merges ecotoxicology, microbiology, and environmental policy.

2. Physicochemical properties and bioavailability

MPs and NPs are distinguished from traditional pollutants by their complex and heterogeneous physicochemical profiles, which profoundly influence their environmental behavior and biological interactions. Key characteristics, including particle size, morphology, polymeric composition, degree of environmental weathering, and surface functionalization, play critical roles in governing their transport dynamics, partitioning, and interaction with biota in aquatic environments [27, 28]. These parameters collectively modulate their bioavailability, toxicokinetics, and propensity to act as carriers for co-contaminants, thereby shaping their ecotoxicological footprint. A mechanistic understanding of these properties is, therefore, fundamental for accurately predicting their environmental fate and assessing their risk potential within freshwater, estuarine, and marine ecosystems.

2.1 Size, shape, polymer type, and surface characteristics

The environmental behavior and biological interactions of MPs and NPs are intrinsically governed by their physical dimensions and molecular architectures, which collectively shape their mobility, reactivity, and toxicological profiles within aquatic systems [29]. Among these, particle size is a critical determinant of uptake pathways, cellular internalization mechanisms, and systemic distribution across biological compartments [29] (**Figure 1**). MPs (>1 μm to <5 mm) are predominantly ingested by larger aquatic organisms through traditional feeding behaviors, whereas NPs (<1 μm) possess the capacity to translocate across biological membranes via

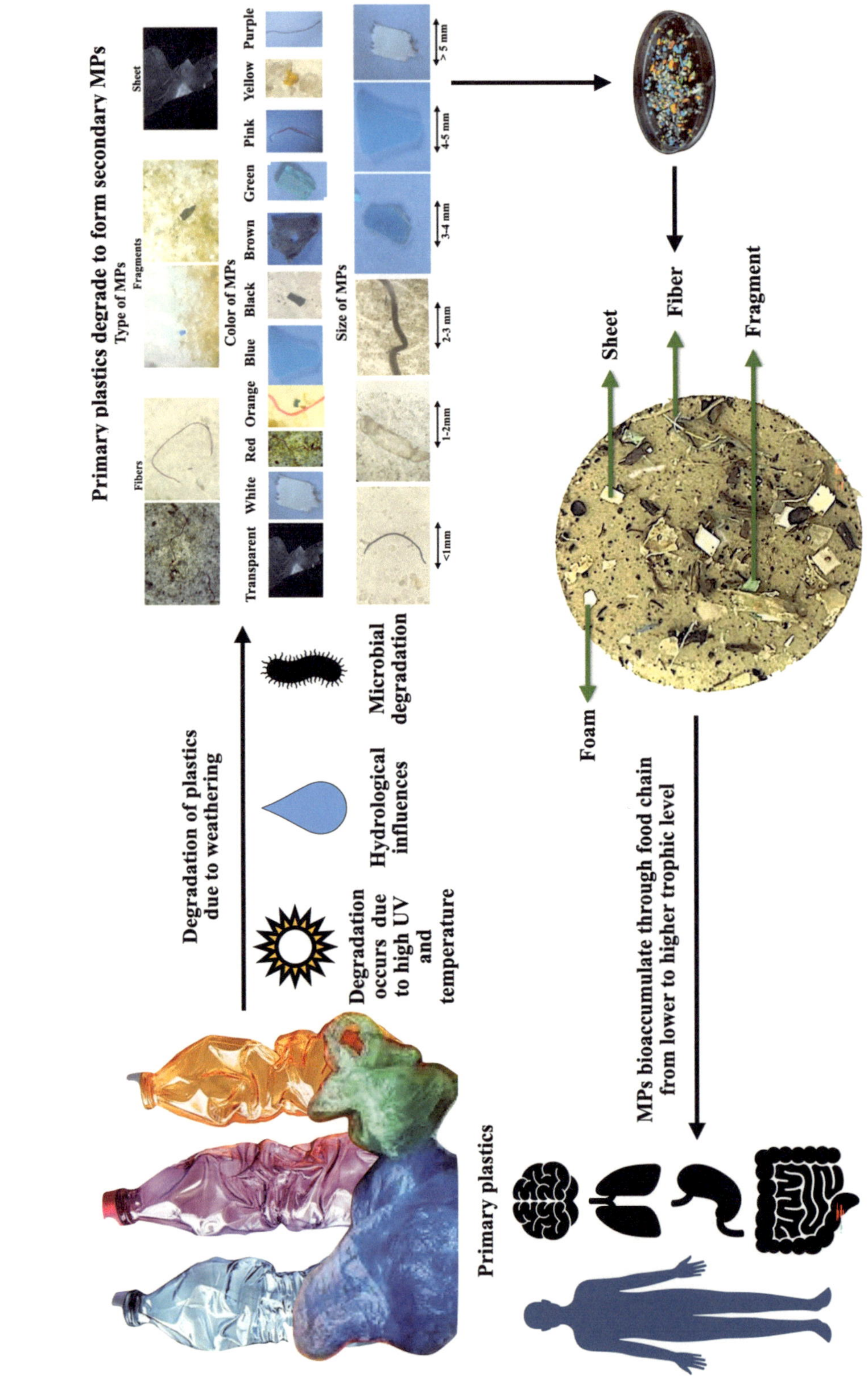

Figure 1.
Pathways of plastic degradation and microplastic formation with bioaccumulation.

endocytosis, passive diffusion, or paracellular transport routes [30]. The elevated surface area-to-volume ratio inherent to smaller particles augments their chemical reactivity and enhances their capacity to sorb environmental contaminants [31]. Particle shape – whether fibrous, fragmented, spherical, film-like, or foamed – plays a pivotal role in determining the likelihood of ingestion and the extent of mechanical injury inflicted upon organisms [28, 32]. For instance, fibrous MPs are more prone to entrapment within respiratory or gastrointestinal tissues, whereas angular fragments may cause epithelial abrasion and localized inflammation [28, 32]. The polymer composition, encompassing materials such as polyethylene (PE), polypropylene (PP), polystyrene (PS), polyethylene terephthalate (PET), and polyvinyl chloride (PVC), dictates intrinsic properties such as hydrophobicity, density, chemical resilience, and affinity for co-contaminants [33]. Buoyancy behavior, which is closely linked to polymer density, determines vertical distribution in the water column: low-density polymers (e.g., PE, PP) tend to remain in surface waters and contribute to neustonic aggregations, whereas high-density polymers (e.g., PET, PVC) are more likely to sediment and contribute to benthic accumulation [34]. Surface characteristics, including textural roughness, electrostatic charge, and the presence of functional chemical groups, undergo dynamic transformations during environmental weathering via photodegradation, mechanical erosion, and biofouling [35]. These processes not only alter the physicochemical reactivity of the plastic particles but also modulate microbial colonization patterns and contaminant-binding affinities [35]. Aged or weathered particles often develop cracked or pitted morphologies that significantly increase sorptive capacity relative to their pristine analogs, thereby enhancing their role as vectors for persistent pollutants and microbial assemblages [36]. Collectively, the diversity and variability of these attributes introduce significant heterogeneity into the environmental dynamics of MPs and NPs. This complexity poses substantial challenges for accurately modeling their fate, transport, exposure routes, and associated ecotoxicological risks in aquatic ecosystems [37].

2.2 Adsorption of environmental pollutants and additives

MPs and NPs exhibit a remarkable ability to adsorb and concentrate a broad spectrum of environmental contaminants due to their hydrophobic surfaces and high specific surface area [38]. These include POPs such as PCBs, PAHs, and dioxins; heavy metals (e.g., lead, cadmium, mercury); pharmaceuticals and personal care products (PPCPs); antibiotics and antimicrobial agents; and plastic additives (e.g., phthalates, bisphenol A, flame retardants) [39, 40]. This sorptive behavior transforms MPs/NPs into vectors or "Trojan horses" for pollutant delivery to aquatic organisms. The adsorption-desorption dynamics are governed by various factors, including polymer type, surface aging, environmental pH, salinity, and the presence of natural organic matter [36]. For instance, weathered polystyrene microbeads show enhanced binding to hydrophobic pollutants like phenanthrene and fluoranthene, which can then be transferred to fish and bivalves upon ingestion [41, 42]. In controlled exposure studies, organisms that ingested contaminated plastics exhibited higher tissue concentrations of these pollutants compared to those exposed to the contaminants alone [43]. Additionally, the release of leachable additives from plastics, such as UV stabilizers and endocrine-disrupting compounds, further compounds the toxicological burden. These chemicals may disrupt hormone signaling, impair immune function, and alter reproductive physiology in aquatic biota, even at sublethal concentrations [42, 43].

2.3 Influence on environmental fate and bioavailability

The bioavailability of MPs and NPs, defined as the extent to which these particles can be absorbed by living organisms, is closely governed by their intrinsic physicochemical properties and the dynamic environmental contexts in which they reside [44]. Particle size and density are primary determinants: smaller, neutrally buoyant particles are more likely to remain suspended in the water column for extended periods, thereby increasing the likelihood of trophic exposure across multiple ecological levels [45]. Aggregation and sedimentation, often driven by surface fouling through biofilm formation and interactions with dissolved organic matter, alter particle density and promote vertical migration and deposition into benthic substrates [1, 46]. Additionally, environmental partitioning processes allow MPs and NPs to transition between surface waters, sediment layers, and biological compartments depending on local hydrodynamic forces, turbulence regimes, and salinity gradients [1, 46].

Once internalized by organisms, particle-specific traits further influence cellular uptake and systemic translocation. For instance, positively charged nanoplastics have demonstrated enhanced electrostatic interaction with negatively charged cell membranes, thereby promoting endocytic uptake and intracellular accumulation [47]. Experimental evidence has documented the ability of polystyrene nanoplastics to traverse the intestinal epithelium and accumulate in sensitive tissues such as the liver, brain, and gonads of fish [18, 48, 49]. Notably, aged or environmentally weathered plastics display elevated bioavailability and toxicity due to oxidative surface modification and increased affinity for biomolecular interactions [50]. These surface alterations may cause plastics to mimic natural prey particles, elevating the risk of mistaken ingestion by filter feeders, detritivores, and planktivorous species [18, 48, 50].

Moreover, the synergistic combination of bioavailable plastic particles with adsorbed environmental toxicants, such as heavy metals, polycyclic aromatic hydrocarbons (PAHs), or POPs, results in multifaceted exposure scenarios that are not adequately captured by conventional ecotoxicological models [37]. This complexity underscores the urgent need for incorporating comprehensive physicochemical profiling of plastic particles into exposure assessment protocols and ecological risk frameworks, ensuring a more accurate prediction of their biological impacts across ecosystems.

A major bottleneck in advancing ecological and health risk assessment is the inconsistent analytical pipeline used across studies. For microplastics, Fourier-transform infrared (FTIR) spectroscopy and Raman spectroscopy are the most widely adopted methods for polymer identification; however, differences in sampling, filtration, chemical digestion, and spectral library matching produce large inter-study variability [18, 48]. For nanoplastics (<1 µm), the situation is worse: validated, high-throughput detection in biological tissues is in its infancy. To improve comparability, we recommend (a) reporting particle counts by size-bin (e.g., <1 µm, 1–10 µm, 10–100 µm, ><100 µm), standardized protocols are urgently needed to reduce variability and improve comparability across studies. These protocols should include: (a) reporting particle size distribution and mass when possible; (b) always declaring digestion and density-separation methods; (c) including spectral match scores for polymer identification; and (d) reporting limits of detection (LODs) for each matrix. Standard reference materials and interlaboratory calibration exercises should be prioritized by the research community and journal editors alike.

3. Ecotoxicological effects on aquatic biota

The rising presence of MPs and NPs in aquatic ecosystems has catalyzed a paradigm shift in ecotoxicology. No longer viewed as inert particles, these synthetic polymers, both pristine and weathered, exhibit diverse and often deleterious biological effects across taxa [51]. The toxicological footprint of MPs and NPs spans cellular, organismal, and population levels, interfering with physiological homeostasis, reproductive success, and ecological integrity [52]. Their effects are mediated through a combination of physical damage, chemical leaching, adsorbed toxicants, and biological interactions, including immune and microbiome disruptions [52]. This chapter synthesizes key mechanistic insights and experimental evidence to critically evaluate the multiscale ecotoxicological impacts of MPs and NPs on aquatic life.

3.1 Cellular and subcellular mechanisms of toxicity

MPs and NPs affect cells through physical interactions (membrane perturbation, particle-induced abrasion), chemical leaching (additives and sorbed toxicants), and indirect effects via the plastisphere (pathogen exposure and microbial metabolites) [53]. The predominant mechanistic cascade observed across taxa begins with particle contact and uptake (phagocytosis/endocytosis for NPs), followed by oxidative stress (mitochondrial dysfunction and ROS generation), inflammatory signaling pathways (elevated cytokines such as IL-1β and TNF-α), genotoxic stress (DNA strand breaks, micronuclei formation), and, in chronic contexts, altered cell proliferation and apoptosis via caspase cascade activation [53]. The severity of these pathways is strongly modulated by particle dose, chemical load (adsorbed contaminants), and the organismal capacity for antioxidant defense.

3.1.1 Oxidative stress and ROS generation

One of the most consistent and well-characterized responses to plastic particle exposure is the overproduction of ROS [54]. Upon internalization, MPs and NPs interfere with mitochondrial function and redox balance [55], leading to lipid peroxidation, protein carbonylation [56], DNA strand breaks, and base modifications [56]. This oxidative imbalance compromises membrane integrity and damages key cellular macromolecules, initiating cytotoxic and mutagenic processes [57]. Several studies on zebrafish (*Danio rerio*), mussels (*Mytilus galloprovincialis*), and crustaceans have demonstrated significantly elevated levels of malondialdehyde (MDA) and reduced antioxidant enzyme activities (e.g., superoxide dismutase, catalase, glutathione peroxidase) following MP/NP exposure [58–60].

3.1.2 Inflammation, apoptosis, and genotoxicity

Chronic exposure to MPs and NPs initiates inflammatory signaling cascades, marked by the upregulation of pro-inflammatory cytokines (e.g., IL-1β, TNF-α) [54, 61] and the recruitment of immune cells [61]. In aquatic invertebrates and fish, this inflammatory state is often linked to tissue degeneration in the gut, liver, and gills [62]. Moreover, persistent cellular stress induces apoptosis through mitochondrial (intrinsic) and death receptor (extrinsic) pathways [63]. Key markers, such as Bax/Bcl-2 ratios and caspase-3 activation, have been observed in exposed organisms, indicating programmed cell death as a primary toxic response [64]. Genotoxic effects are also well-documented. Comet assays and micronucleus tests have shown DNA fragmentation and chromosomal aberrations in cells exposed to both MPs and associated contaminants [65]. In bivalves, DNA damage correlates strongly with plastic particle concentration and exposure duration [66].

3.2 Organismal-level effects

Beyond cellular toxicity, MPs and NPs disrupt whole-organism physiology and behavior, with significant implications for growth, development, fitness, and survival. These effects are often dose-dependent but can also occur at environmentally relevant concentrations [67]. At the organ level, repeated ingestion of MPs leads to gut epithelial inflammation, altered nutrient assimilation, and changes to the gut microbiome composition. In fish, nanoplastics can translocate to the liver and brain, invoking hepatic lipid accumulation and neuroinflammatory changes associated with behavioral deficits [54]. Filter-feeders accumulate large particle loads in the digestive tract, which can reduce filtration efficiency and growth rates and, for commercially important species, create direct food-safety concerns when consumed whole.

3.2.1 Impaired reproduction and development

Reproductive toxicity is one of the most concerning outcomes of plastic exposure in aquatic fauna [68]. Studies on model species such as zebrafish and *Daphnia magna* have demonstrated reduced fecundity and egg viability, decreased sperm motility, altered oocyte morphology, and developmental abnormalities such as spinal deformities, yolk sac edema, and cardiac dysfunction in embryos [69–71]. Mechanistically, these effects are often linked to endocrine disruption via plastic additives (e.g., bisphenol A, phthalates), oxidative stress in reproductive tissues, and interference with hormonal signaling pathways (e.g., estrogen receptor modulation) [72].

3.2.2 Neurotoxicity and behavioral disruptions

MPs and NPs have been shown to penetrate the blood-brain barrier in fish, leading to neuroinflammation, oxidative stress, and neuronal apoptosis [73]. Resultant behavioral impairments include altered predator avoidance, reduced locomotion and foraging efficiency, cognitive deficits, and anxiety-like responses [73]. Behavioral endpoints are increasingly recognized as sensitive biomarkers of sublethal toxicity, reflecting both neurological and systemic stress [74]. In aquatic invertebrates like *Artemia* and *Daphnia*, exposure to polystyrene NPs has been associated with erratic swimming and impaired escape responses – effects that compromise ecological fitness and survival [75].

3.2.3 Immune and endocrine modulation

Plastic particle exposure can dysregulate immune function through either immunosuppression or hyperactivation [76]. In fish and mollusks, altered expression of immune-related genes (e.g., lysozyme, C-type lectins) has been documented, along with increased susceptibility to opportunistic infections [77]. Endocrine disruption, a hallmark of plastic additive toxicity, has been observed through changes in gonadal development and sex hormone levels, vitellogenin expression (a biomarker for estrogenic activity), and hypothalamic-pituitary-gonadal (HPG) axis signaling [68]. These disruptions have long-term consequences for population dynamics, especially in species with short generation times or high reproductive sensitivity.

3.3 Species-specific sensitivity: From plankton to fish

The ecotoxicological effects of MPs and NPs are not uniformly expressed across aquatic taxa [78]. Instead, species-specific sensitivity is governed by a combination of physiological traits, feeding strategies, habitat preferences, and life stage vulnerabilities [79]. A comprehensive understanding of these differences is essential for developing ecological risk models and prioritizing conservation strategies.

3.3.1 Planktonic organisms: The gatekeepers of aquatic food webs

Phytoplankton and zooplankton represent critical basal components of aquatic ecosystems, and their susceptibility to MPs/NPs poses a direct threat to trophic stability [80]. Phytoplankton, such as *Skeletonema costatum* and *Thalassiosira pseudonana,* exhibit reduced photosynthetic efficiency and chlorophyll content when exposed to MPs, particularly NPs, which adhere to cell surfaces and interfere with light absorption and gas exchange [80, 81] (**Table 1**). Zooplankton, including copepods (*Calanus helgolandicus*) and cladocerans (*Daphnia magna*), readily ingest microplastics due to their non-selective filter-feeding behavior [82, 83]. This ingestion reduces energy assimilation, alters gut residence time, and impairs fecundity [82]. Due to their small size and high metabolic rate, planktonic organisms also experience bioaccumulation of associated toxicants, making them critical bioindicators and vectors for trophic transfer [82] (**Figure 2**).

3.3.2 Invertebrates: Mollusks and crustaceans

Sessile filter feeders, such as bivalves (e.g., *Mytilus edulis*, *Crassostrea gigas*), are highly sensitive to MPs/NPs due to continuous exposure to suspended particles [84]. Documented effects include impaired filtration rates, histopathological damage in gill and digestive tissues, hemocyte dysfunction, and oxidative stress [84]. Crustaceans, such as amphipods and shrimp (*Palaemonetes pugio*), display sublethal effects like reduced molting success, altered locomotor activity, and endocrine disruption following chronic exposure [85].

Organism group	Representative Sp.	Feeding mode	Exposure route	Observed effects	Trophic transfer potential	Ref.
Phytoplankton	*Skeletonema costatum*, *T. pseudonana*	Autotrophic	Direct suspension contact	Reduced photosynthesis, cell aggregation, pigment loss.	Low (base of the food web, no ingestion)	[80, 81]
Zooplankton	*Daphnia magna*, *Calanus helgolandicus*	Filter feeding	Ingestion of suspended MPs/NPs	Gut blockage, reduced fecundity, impaired mobility.	High (key vector to fish, invertebrates).	[82, 83]
Bivalves	*Mytilus edulis*, *Crassostrea gigas*	Filter feeding	Ingestion via filtration	Inflammation, digestive tissue damage, immune suppression.	Medium–High (bioaccumulated in edible tissues)	[84]
Crustaceans	*Palaemonetes pugio*, *Gammarus spp.*	Omnivorous/ Detritivorous	Ingestion of contaminated detritus	Molting disruption, oxidative stress, behavioral alterations.	Medium (preyed on by fish and birds)	[85]
Fish larvae/ Juveniles	*Danio rerio*, O. niloticus	Active predation	Ingestion via prey/ particles	Reduced growth, developmental defects, gut inflammation.	High (critical developmental stage sensitivity)	[86]
Fish – adults	*Dicentrarchus labrax*, *Oncorhynchus mykiss*	Carnivorous	Prey consumption, direct ingestion	Liver damage, neurotoxicity, reproductive failure.	High (consumed by humans and predators).	[87, 88]
Birds and marine mammals	*Larus spp.*, *Phocoena phocoena*	Piscivorous	Ingestion of contaminated fish	Gut blockage, reduced nutrient absorption, immune suppression.	High (top predators, plastic accumulation)	[89–91]
Humans		Omnivorous	Seafood consumption, water intake	Under research: inflammation, endocrine disruption, oxidative stress (preliminary).	Terminal consumer, a significant concern for food safety.	[91]

Table 1.
Species-specific sensitivities and trophic transfer pathways of MPs and NPs in aquatic ecosystems.

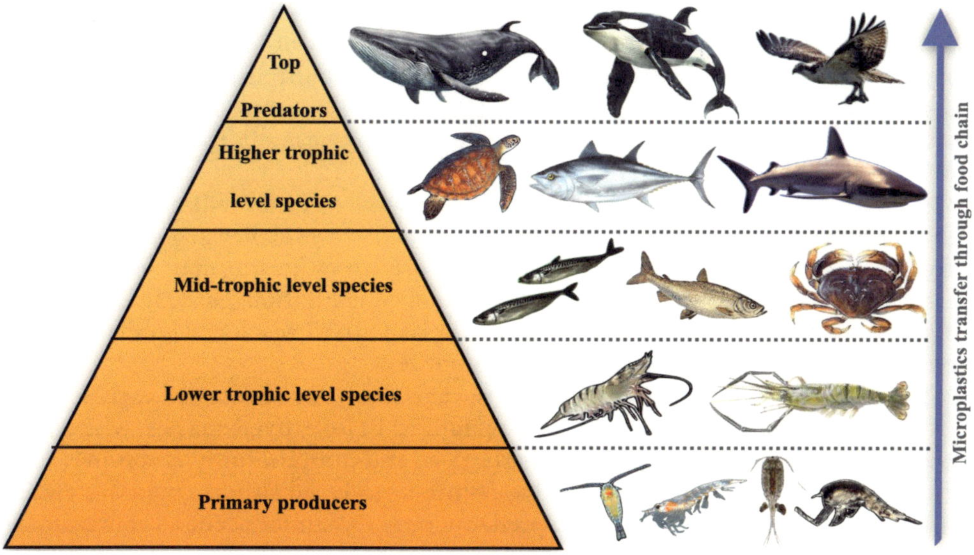

Figure 2.
Schematic of trophic transfer in aquatic food webs

.

3.3.3 Fish: Complexity across life stages

Fish represent the top vertebrate consumers in many aquatic systems and exhibit complex responses depending on age, species, and trophic level [92]. Laboratory studies have shown that larvae and juveniles are more vulnerable due to underdeveloped detoxification systems and higher surface-to-volume ratios [93–95]. Gastrointestinal accumulation of MPs is well-documented, with retention times ranging from hours to days depending on particle size and composition [96]. Organ-level impacts include liver inflammation, gut microbiota dysbiosis, and neurotoxic effects, particularly in species like *Danio rerio* (zebrafish) and *Oreochromis niloticus* (Nile tilapia) [86]. Reproductive endpoints, such as reduced gamete quality and altered hormone profiles, are frequently reported in chronic exposure studies [87, 88]. Importantly, predatory species that consume smaller contaminated organisms may experience compounded effects through biomagnification, although data are still emerging [94].

3.3.4 Inter-species variability and research gaps

Differences in MP/NP effects across species are influenced by feeding mode (filter feeder, deposit feeder, predator), particle selectivity, habitat exposure (benthic vs. pelagic), and detoxification and excretion mechanisms [97]. Current ecotoxicological frameworks often fail to account for such ecological diversity [98]. Future studies must adopt multi-species testing protocols under environmentally relevant conditions to accurately map sensitivity landscapes.

3.4 Trophic transfer and food web implications

The ecological implications of plastic pollution extend far beyond individual toxicity. MPs and NPs are now recognized as vectors for trophic transfer, whereby ingested particles pass from prey to predator, carrying both physical particles and sorbed pollutants up the food web [52]. Trophic transfer of MPs/NPs has been confirmed in both laboratory and field studies across multiple aquatic food chains (zooplankton to small fish to predators). For instance, polystyrene MPs have been shown to move from *Artemia to Pagrus* major, accumulating in the liver and intestines [89, 90]. Bivalves and crustaceans to fish and birds, where mussels consumed by demersal fish have been found to serve as intermediate reservoirs of MPs, and fish to marine mammals and humans, where evidence suggests MPs accumulate in the gastrointestinal tracts of top predators, raising concerns for higher vertebrates, including humans, through seafood consumption [90, 91]. These pathways are facilitated by the persistence of plastic particles, low degradation rates, and limited excretion efficiency in many organisms.

Trophic transfer introduces systemic risks to ecosystem function, such as bioaccumulation of MPs and associated pollutants in long-lived species, which could result in chronic toxicity, hormonal disruption, and immunosuppression [41]. Loss of keystone species due to reproductive failure or reduced fitness could destabilize community structures [99]. Altered predator-prey dynamics occur as contaminated prey exhibit reduced escape behaviors or survival rates [100]. Energy transfer disruption, where ingestion of inert plastic particles leads to reduced nutritional intake and impaired growth across trophic levels, is another significant concern [101]. Such changes threaten the resilience and productivity of aquatic ecosystems, particularly in biodiverse and economically important habitats like coral reefs, estuaries, and mangroves [102].

The cascading effects of trophic transfer can influence ecosystem services such as fisheries yield, water purification, and carbon sequestration [104]. Moreover, the intersection with human food webs, particularly through the consumption of shellfish and small pelagic fish, has introduced a direct route of microplastic exposure to humans, sparking global concern over food safety and health [52].

4. Human health implications

The pervasiveness of MPs and NPs in aquatic ecosystems has escalated concerns over their potential transference from the environment to human populations [52]. As these synthetic particles integrate into food chains and water systems, they emerge not only as environmental pollutants but also as emerging contaminants of global public health concern [103]. This chapter critically evaluates the established and emerging pathways of human exposure, explores toxicological evidence from experimental models, and examines the potential health outcomes associated with chronic and low-dose plastic particle exposure (**Figure 3**).

4.1 Pathways of human exposure

Human exposure arises from three principal routes: ingestion (seafood, salt, water), inhalation (airborne fallout, indoor dust), and dermal/occupational contact (workplace aerosolization, recycling shops). Quantitatively, ingestion through seafood remains the most studied route for integrated body burden; however, inhalation is gaining evidence for contributing non-trivial particle mass due to

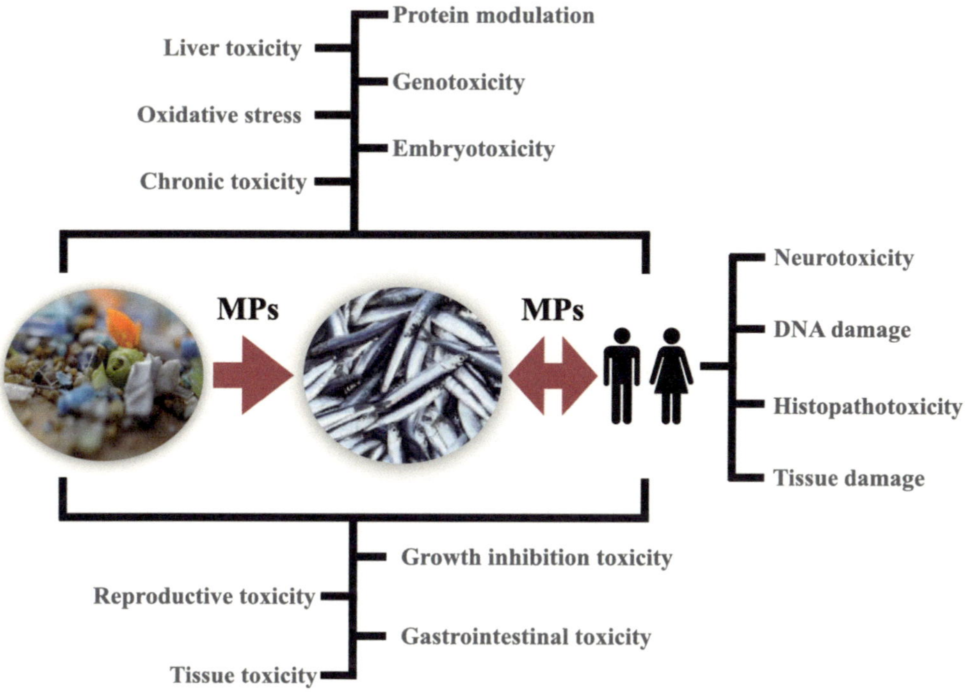

Figure 3.
Toxicological effects of microplastics on aquatic biota and humans.

indoor dust loading and synthetic textile abrasion. For each route, we recommend reporting both particle number and mass per unit intake, as well as context-specific physicochemical descriptors (size, polymer type).

4.1.1 Seafood and aquatic products

Among the most direct and concerning routes of human exposure to MPs/NPs is through the consumption of seafood, particularly filter-feeding organisms such as bivalves (e.g., Mytilus spp.), crustaceans, and small pelagic fish (e.g., Sardina pilchardus) [82, 104] (**Table 2**). These species are often consumed whole or partially whole, increasing the likelihood of ingesting MPs residing in their digestive tracts [105]. Studies have documented tens to hundreds of MP particles per individual mussel or oyster, particularly in urban coastal regions [106]. NPs pose an even greater risk due to their ability to cross biological barriers, potentially entering edible tissues [107]. The bioaccumulation of plastic-associated contaminants such as bisphenol A (BPA), phthalates, and POPs exacerbates toxic potential [108, 109]. Human exposure through seafood is not merely incidental but bioavailable and biologically relevant, particularly in high seafood-consuming populations [110].

4.1.2 Drinking water and atmospheric deposition

MPs/NPs have been widely detected in both tap and bottled drinking water, with concentrations ranging from a few to thousands of particles per liter, depending on

Exposure pathway	Major sources	Representative evidence	Potential health impact	Ref.
Dietary – seafood and aquatic products	Bivalves, crustaceans, fish (especially consumed whole).	MPs found in 100% of commercial mussels/oysters in several studies.	Inflammation, endocrine disruption, oxidative stress	[104, 105]
Drinking water	Bottled water (PET), tap water, groundwater	Bottled water: 10–10,000 MPs/L; tap water (global average): 5–20 MPs/L.	Gut irritation, tissue accumulation, and immune modulation.	[111]
Airborne inhalation	Indoor dust, synthetic textiles, urban air fallout.	Indoor air: up to 10,000 MPs/day inhaled per person in industrial zones.	Respiratory inflammation, allergic response, systemic exposure.	[112]
Occupational exposure	Plastic manufacturing, recycling, synthetic textiles, construction.	Workers show elevated markers of respiratory and dermal inflammation.	Chronic lung disease, skin disorders, reproductive toxicity.	[112]
Recreational activities	Swimming, beach activities, and water sports in polluted environments	Exposure via ingestion/contact with contaminated water or sand.	Unknown; assumed minor but chronic exposure risk.	[113]
Transplacental exposure	Placental tissue, umbilical cord (emerging data).	MPs detected in human placentas.	Potential developmental disruption, fetal immune priming.	[72, 114]
Food contamination (non-seafood)	Salt, sugar, milk, honey, vegetables (from plastic packaging/processing).	MPs found in table salt (~200–500 particles/kg), teabags, and dairy products.	Chronic ingestion, endocrine disruption risk	[115]

Table 2.
Summary of human exposure pathways to MPs and NPs, evidentiary support, and health implications.

the region, treatment methods, and packaging materials [111]. Bottled water, particularly in PET containers, often contains fragmented plastic particles released during storage and transport [111]. Municipal tap water systems, even in developed nations, are not immune, particularly when relying on surface water sources [116]. Moreover, airborne microplastic fallout has been recorded in indoor and outdoor environments, suggesting inhalation and mucosal exposure as a relevant route, especially in urban centers and industrial settings [112]. These particles can be deposited in the respiratory tract and swallowed, making this a dual exposure route [117].

4.1.3 Occupational and recreational exposure

Populations engaged in plastic manufacturing, recycling, and textile industries face disproportionately high MP/NP exposure through inhalation and dermal contact [118]. Workers in plastic molding facilities exhibit elevated inflammatory markers and respiratory symptoms [119]. Recreational exposure (e.g., swimming, beach sports) in contaminated water bodies may also lead to incidental ingestion or skin contact [113]. Although less documented, this pathway represents a chronic, low-dose exposure scenario for millions globally.

4.2 Toxicological evidence in mammalian models and in vitro studies

Due to ethical constraints in human testing, much of the mechanistic understanding of plastic toxicity stems from mammalian in vivo models and cell-based in vitro assays [120]. Rodent studies reveal systemic uptake of NPs following oral ingestion, with particles found in the liver, spleen, kidneys, and even the brain [121]. In mice, polystyrene NPs induce gut inflammation, altered gut microbiota composition, and metabolic dysfunction [122]. In vitro studies with human lung and intestinal epithelial cells show increased ROS production, membrane disruption, and DNA damage upon plastic exposure [123]. These results provide compelling evidence that MPs and NPs are bioactive agents capable of eliciting toxicological responses at multiple biological levels. We currently lack robust, reproducible toxicokinetic models for plastics in humans. Key unresolved parameters include the absorption fraction across intestinal epithelia for polydisperse particles, distribution half-lives in organs, metabolic transformation (if any), and clearance mechanisms (urinary vs. biliary excretion). Advancing these models requires coordinated animal-to-human scaling studies, validated particle tracers, and harmonized reporting of dose metrics (mass, particle count, surface area).

4.3 Potential health outcomes

To move from plausibility to causality, prospective cohort studies with archived biological samples (blood, stool, cord blood) and repeated exposure measures are required. A recommended study design: a birth cohort with maternal exposure assessment (dietary questionnaires and measurement of MPs in maternal blood/placenta), infant developmental follow-ups (neurodevelopmental milestones, immune function, growth metrics), and nested case–control analyses for outcomes such as autoimmune disorders, metabolic syndromes, and reproductive health.

4.3.1 Inflammation and immune dysregulation

The primary physiological response to MPs/NPs is chronic inflammation. Phagocytic immune cells attempt to engulf plastic particles, leading to a frustrated phagocytosis response, ROS over production, and cytokine release [124]. Macrophage activation, complement system perturbation, and increased levels of pro-inflammatory interleukins (e.g., IL-6, TNF-α) have been observed in murine and human cell studies [125]. Chronic exposure could predispose individuals to autoimmune conditions, allergies, and systemic inflammatory disorders.

4.3.2 Endocrine disruption and reproductive risk

Plastic particles can act as endocrine-disrupting chemical (EDC) carriers, interfering with hormone signaling and reproductive functions [72]. Exposure to MPs coated with nonylphenol and phthalates has been linked to altered testosterone levels, reduced sperm quality, and disrupted estrous cycles in rats [68]. Human epidemiological studies are emerging, with correlations between higher urinary BPA levels and reproductive dysfunction, though causality remains to be established [114].

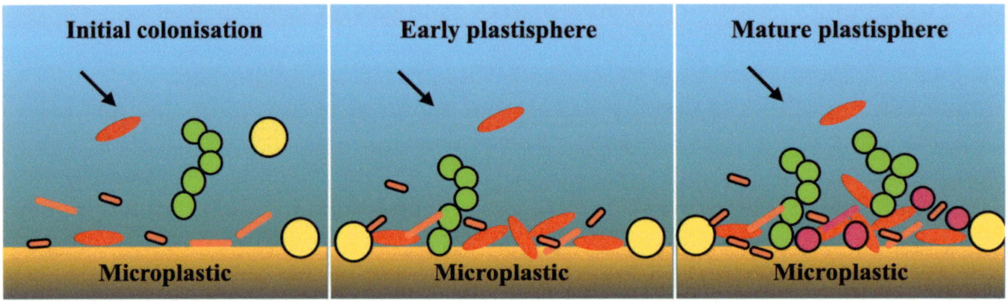

Figure 4.
Stages of microbial biofilm development on microplastic surfaces.

4.3.3 Neurotoxicity and gut microbiome alterations

NPs may cross the blood–brain barrier, leading to neuroinflammatory responses and cognitive impairments [126]. Animal studies demonstrate microglial activation, oxidative stress in brain tissues, and impaired learning behavior in plastic-exposed rodents [127–129]. Simultaneously, gut-brain axis disruption has been observed, with significant dysbiosis of the intestinal microbiome, reduced microbial diversity, and increased leaky gut syndrome [130]. These changes are associated with mood disorders, altered metabolism, and heightened systemic vulnerability to disease.

5. The plastisphere: Microbial colonization of plastic surfaces

Plastics in aquatic environments have rapidly evolved from inert pollutants to dynamic microbial ecosystems [131]. The "plastisphere" refers to biofilm communities that colonize plastic surfaces, distinct from those found on natural substrates [132] (**Figure 4**). These communities foster unique ecological niches, drive biogeochemical cycling, and serve as vectors for pathogens and antimicrobial resistance genes (ARGs) [132]. This chapter delves into the mechanisms of microbial colonization, ecological implications, and the public health threats posed by the plastisphere, framing plastic debris not merely as waste but as mobile microbial habitats with profound consequences for ecosystem integrity and disease transmission.

5.1 Formation and succession of biofilms on plastics

The colonization of plastics begins almost immediately upon contact with aquatic environments [133]. This process follows a structured biofilm formation cascade, which includes the adsorption of organic molecules (conditioning film), the attachment of pioneer microbial taxa, the maturation of biofilm through quorum sensing and EPS (extracellular polymeric substances) secretion, and detachment and dispersal [134, 135]. Unlike natural substrates (e.g., wood, sediment), plastics provide hydrophobic, persistent, and often nutrient-scarce surfaces that select for specialized microbial consortia [136]. Early colonizers often include proteobacteria, bacteroidetes, and cyanobacteria, later succeeded by fungi, archaea, diatoms, and occasionally macroorganisms such as larval invertebrates [137]. Recent omics-based studies (metagenomics, 16S rRNA sequencing) show that plastisphere biofilms differ taxonomically and functionally from surrounding water and sediment microbiomes [138].

5.2 Ecological role of the plastisphere in aquatic systems

While often perceived solely as an anthropogenic disturbance, the plastisphere has evolved into a complex ecological niche with multifaceted functions. Microbial communities inhabiting plastic surfaces actively participate in nutrient cycling by metabolizing nitrogen, sulfur, and carbon compounds, thereby integrating into key biogeochemical processes [139]. These artificial substrates also serve as habitats for extremophile microorganisms capable of thriving on nutrient-poor, UV-exposed, and chemically contaminated plastic surfaces due to their remarkable metabolic flexibility [140]. Furthermore, plastics act as effective vectors for the long-distance dispersal of microbes and sessile organisms, facilitating the spread of invasive species and potentially destabilizing local biodiversity and ecosystem function [141]. Over time, plastics undergo microbial succession, with shifts in colonizing taxa often influenced by biofilm thickness and environmental variables such as temperature, salinity, and UV exposure [142]. Consequently, plastics function as novel benthic and pelagic habitats, reshaping microbial biogeography across both freshwater and marine ecosystems [143].

5.3 Plastics as vectors for pathogens and antibiotic resistance

Perhaps the most concerning facet of the plastisphere is its ability to harbor and disseminate pathogens and ARGs [144]. Numerous studies have identified opportunistic human and animal pathogens (e.g., Vibrio spp., Aeromonas, Pseudomonas, Escherichia coli) colonizing marine plastics [145–147]. The plastisphere enhances pathogen survival by providing physical protection from predation, desiccation, and UV exposure [148]. More alarmingly, plastics serve as hotspots for horizontal gene transfer (HGT): ARGs, such as those conferring resistance to β-lactams, tetracyclines, and sulfonamides, are frequently detected on plastic biofilms [144]. Dense microbial consortia in biofilms enhance HGT via conjugation, transformation, and phage-mediated transduction [149]. In aquatic environments where plastics intersect with effluents (e.g., hospital wastewater, urban runoff), this leads to a perfect storm of gene exchange, resistance amplification, and global dissemination [150].

5.4 Implications for disease emergence and environmental AMR

The plastisphere represents an under-recognized yet potent interface between environmental microbiology and public health [133]. One of its most concerning implications is its role as a reservoir for antimicrobial resistance (AMR), where plastics function as long-term carriers and vectors for resistance genes, bridging clinical, environmental, and agricultural domains [151]. Additionally, the plastisphere may contribute to the emergence or re-emergence of waterborne infectious diseases (EIDs), particularly in eutrophic, warming, and polluted aquatic systems [152]. Aquatic organisms that interact with pathogen-contaminated plastics can serve as intermediate hosts for zoonotic pathogens, posing significant risks to fisheries, aquaculture, and food safety [153]. Despite growing scientific evidence, current AMR action plans and water quality monitoring frameworks largely overlook the plastisphere's role. Effectively addressing these challenges necessitates a One Health approach integrating environmental surveillance, microbial risk assessment, and comprehensive regulatory reforms targeting plastic production, use, and wastewater discharge [154].

6. Knowledge gaps and research priorities

Despite the exponential rise in research on MPs and NPs, major knowledge gaps persist in understanding their toxicological impacts, human health risks, and environmental dynamics. Current models lack ecological realism, human exposure pathways remain under-characterized, and interactions with co-contaminants and microbiomes are poorly resolved [155]. This chapter critically examines the key limitations in existing methodologies and proposes a roadmap for future research. The integration of multidisciplinary approaches, standardized protocols, and next-generation analytical tools is essential for improving risk assessment and shaping informed policy. Bridging these gaps will enable a holistic, predictive framework for plastic pollution management in the Anthropocene [156].

6.1 Limitations in current ecotoxicological models

Ecotoxicological assessments of MPs and NPs are often constrained by oversimplified laboratory models that fail to reflect environmental realities. Most studies use pristine, spherical particles at unrealistically high concentrations and rely on single-species testing, which inadequately captures the complexity of natural ecosystems [157]. In real-world conditions, plastic particles are weathered, chemically altered, and highly heterogeneous in terms of size, shape, polymer composition, and density. They often co-occur with contaminants like heavy metals, hydrophobic organic compounds, and microbial biofilms, leading to multifaceted toxicity profiles. A major limitation is the scarcity of chronic, low-dose exposure studies that mirror environmentally relevant scenarios. Moreover, the taxonomic range is narrow; ecotoxicity data are predominantly based on standard model organisms such as *Daphnia magna* and zebrafish, offering limited insight into impacts on ecologically critical species [158]. Community and ecosystem-level endpoints such as biodiversity, trophic interactions, and nutrient cycling are significantly underrepresented in current assessments. Additionally, nanoplastics (<100 nm) pose unique toxicokinetic challenges due to their capacity to penetrate cellular membranes and cross biological barriers, yet they remain poorly characterized due to methodological and analytical limitations [159]. Addressing these gaps is essential for developing accurate risk assessment frameworks and informing environmental policy.

6.2 Challenges in human risk assessment

Human exposure models to MPs and NPs are currently hindered by significant scientific uncertainties. Detection and quantification methodologies, particularly for nanoplastics in biological matrices, remain inconsistent and lack standardization [160]. Additionally, toxicokinetic data are scarce, and we have limited understanding of the absorption, distribution, metabolism, and excretion (ADME) of MPs and NPs in the human body [159]. No established dose-response thresholds exist for plastic particle exposure across different life stages, further complicating risk assessment [161]. Although MPs have been detected in human placenta, blood, lungs, and stool, direct causal relationships with disease outcomes have yet to be definitively proven [162]. Current health risk assessments also frequently overlook the role of additive leachates, such as phthalates and bisphenols, the potential synergistic toxicity with co-occurring environmental pollutants, and the modulatory influence of the gut microbiome on MP toxicity [162]. Critically, vulnerable populations, including pregnant women, neonates, and individuals with occupational exposure, such as coastal workers, are not systematically considered in existing exposure and regulatory frameworks, representing a major gap in public health protection [163].

6.3 Emerging contaminants and combined effects

Real-world exposure involves complex chemical cocktails. Plastics often adsorb POPs, pharmaceuticals, pesticides, and metals, acting as vectors for combined toxic effects [164]. Key research gaps include interactive effects between plastic particles, associated chemicals, and microbial communities; non-additive toxicological interactions, including antagonism or potentiation; and biotransformation of adsorbed compounds by plastisphere microbes, which may alter bioavailability or toxicity. Moreover, co-occurrence with climate-related stressors (e.g., ocean acidification, temperature rise, hypoxia) remains under-investigated but may amplify plastic toxicity across trophic levels. The field lacks comprehensive mixture toxicity models and cumulative risk assessments that account for real-life exposure scenarios.

6.4 Cross-disciplinary integration and method standardization

A fundamental barrier to scientific progress is the lack of standardized methods across studies. This includes inconsistent sampling, extraction, and quantification protocols; variable definitions of particle size classes (MPs vs. NPs); and the absence of reference materials and validated positive controls for toxicity assays.

To improve reproducibility and comparability, the field needs harmonized guidelines for exposure trials and environmental monitoring, advanced analytical platforms (e.g., FTIR imaging, Raman spectroscopy, FFF-ICP-MS, pyrolysis-GC/MS) that can reliably detect and characterize complex plastic matrices, and cross-validation of *in vitro*, *in vivo*, and *in silico* models. Equally important is fostering interdisciplinary collaboration among marine biologists, chemists, toxicologists, data scientists, and public health experts. Integrated approaches such as systems toxicology, exposomics, One Health, and planetary health frameworks will be key to resolving the complex interplay of plastic pollution across ecosystems and societies.

7. Future perspectives

7.1 Synthesis of key findings

The global burden of MPs and NPs in aquatic environments has become increasingly evident over the last decade, substantiated by multidisciplinary research spanning environmental science, toxicology, microbiology, and human health [165]. The collective body of evidence demonstrates that plastics not only persist across a range of aquatic matrices, from surface waters to deep-sea sediments, but also act as vectors of chemical additives, heavy metals, and microbial pathogens [166]. Physicochemical diversity determines fate and bioavailability; size, shape, surface charge, and polymer type critically influence how MPs and NPs interact with organisms and the environment [159]. These variables modulate the adsorption of POPs and heavy metals, bioavailability to aquatic species, and environmental residence time. Ecotoxicological effects span biological scales, from oxidative stress and genotoxicity at the cellular level to reproductive, behavioral, and immunological impairments in whole organisms. Microplastic toxicity is widespread, species-specific, and exacerbated under combined exposure with other contaminants [94]. Human health implications are increasingly plausible, as evidence from animal models and in vitro systems suggests plastic particles may translocate across biological barriers, induce inflammation, disrupt the endocrine system, and alter the gut microbiome [167–169]. Seafood, drinking water, and inhalation constitute major exposure pathways. The Plastisphere represents a new frontier: the colonization of plastic surfaces by unique microbial communities has led to the emergence of a "plastisphere," a dynamic ecological niche with the potential to disseminate antibiotic resistance genes (ARGs) and serve as reservoirs for pathogenic microbes [143].

7.2 Implications for policy and public health

The transition from scientific understanding to actionable policy is urgently needed to mitigate the impacts of plastic pollution. The complexity of MP interactions across ecological and biological domains necessitates a systems-level, One Health approach that integrates environmental monitoring, public health surveillance, and regulatory reform.

- Standardization of detection and monitoring protocols: To enable comparability across global datasets and establish regulatory thresholds, standardized methodologies for sampling, identification (e.g., FTIR, Raman), and toxicological assessment are essential.

- Precautionary regulations for plastic additives: Many plasticizers and colorants are known endocrine disruptors, yet they remain unregulated in several jurisdictions. Adopting precautionary bans or limits can reduce long-term health risks.

- Inclusion of microplastics in water quality and seafood safety guidelines: National agencies and international bodies (e.g., WHO, FAO) must incorporate microplastic thresholds into food safety and drinking water standards.

- Surveillance of the plastisphere and environmental AMR: Given its role in pathogen and ARG dissemination, the plastisphere must be monitored within existing microbial and AMR surveillance frameworks.

- Public awareness and consumer behavior: Empowering consumers through plastic labeling (e.g., biodegradability, recyclability), promoting behavioral change campaigns, and incentivizing industry reform are key levers in reducing plastic usage.

7.3 Toward sustainable plastic use and pollution prevention

The mitigation of aquatic plastic pollution cannot rely solely on downstream interventions like cleanup technologies or waste management. A transformative shift toward sustainable plastic use is imperative, grounded in the principles of the circular economy, green chemistry, and life-cycle thinking.

- Development of biodegradable and biocompatible polymers: Emerging research in algal-based bioplastics, polyhydroxyalkanoates (PHA), and polylactic acid (PLA) shows promise, though concerns remain regarding degradation products and environmental compatibility.

- Product redesign for deconstruction and reuse: Industrial design must incorporate modularity and end-of-life recyclability into plastic products to facilitate reuse and reduce waste leakage.

- Investment in waste infrastructure and innovation: In many developing economies, the lack of plastic waste collection and treatment infrastructure remains a key bottleneck. Targeted funding and innovation in sorting, pyrolysis, and enzymatic degradation can close this loop.

- International governance and transboundary cooperation: Plastic pollution does not respect borders. As such, binding international frameworks – beyond voluntary initiatives – are necessary. The recently proposed UN Global Plastics Treaty could be instrumental in catalyzing unified global action.

8. Conclusion(s)

The ecotoxicological and human health impacts of MPs and NPs in aquatic environments represent a formidable and multifaceted global challenge. This chapter has critically examined how these synthetic particles – once considered chemically inert – exert complex biological effects across molecular, organismal, ecological, and epidemiological scales. Evidence synthesized across disciplines underscores that MPs and NPs are not passive pollutants but active environmental agents capable of altering cellular function, disrupting ecological interactions, and penetrating human food chains. Their physicochemical diversity, including variations in size, shape, surface chemistry, and polymer type, directly influences bioavailability, contaminant adsorption, and interaction with aquatic organisms. Toxicological manifestations, ranging from oxidative stress, inflammation, and endocrine disruption to neurotoxicity and reproductive failure, have been reported across taxa, from plankton to fish. Additionally, their persistence in trophic networks and capacity for bioaccumulation raise red flags for long-term ecosystem destabilization and reduced biodiversity. Of particular concern is the emerging role of the plastisphere, a complex microbial biofilm community on plastic surfaces that functions as both a vector for pathogens and a reservoir of antimicrobial resistance genes (ARGs). This biofilm-mediated risk bridges environmental and public health domains, complicating water safety, disease emergence, and AMR management under the One Health paradigm. Moreover, the human health dimension, while still in its infancy, suggests tangible risks via dietary, inhalational, transplacental, and occupational exposure routes. Experimental evidence from in vitro and in vivo studies implicates MPs/NPs in gut dysbiosis, immunomodulation, endocrine perturbation, and systemic toxicity. Despite substantial advances, critical knowledge gaps remain. Current ecotoxicological models lack environmental realism, human risk assessments are limited by detection challenges and insufficient toxicokinetic data, and interactions with co-contaminants and climate stressors remain underexplored. Addressing these challenges will require standardized methodologies, robust surveillance systems, interdisciplinary collaboration, and integrative frameworks such as exposomics and systems toxicology. Ultimately, reversing the trajectory of plastic pollution demands more than mitigation; it requires transformational change in how plastics are produced, used, and regulated. This includes the design of biodegradable alternatives, lifecycle product engineering, investment in global waste infrastructure, and the enactment of legally binding international agreements. The future of aquatic ecosystems and human health hinges on our ability to translate scientific insight into decisive policy and sustainable innovation. The window for action is narrowing, but the science is now undeniable.

Acknowledgments

The authors would like to express their sincere gratitude to Professor Dr. R. Rajaram, Research Coordinator, Department of Marine Science, Bharathidasan University, Tiruchirappalli, for his invaluable guidance, mentorship, and academic support throughout the development of this chapter. We extend our heartfelt thanks to our colleagues at the Department of Marine Science for their collaboration and constructive feedback, and to the laboratory and field teams whose assistance was instrumental in executing our research on microplastic pollution. Lastly, the authors acknowledge the unwavering encouragement and moral support of their families throughout the course of this work.

Conflict of Interest

The authors declare no conflict of interest.

DOI: http://dx.doi.org/10.5772/intechopen.1013839

Author details

Sourav Bhattacharya and Rajendran Rajaram*
Bharathidasan University Tiruchirappalli, Tiruchirappalli, India

*Address all correspondence to: rajaramdms@bdu.ac.in

References

[1] Hossain M, 2024. Trending physico-chemical characteristics of microplastics (MPs) and nano-plastics (NPs) in different physical environments and their role in various emerging environmental issues: A review. Available at SSRN 5095223. DOI: 10.2139/ssrn.5095223

[2] Garcia-Muñoz P, Robert D, Ruppert AM, Keller N. Microplastics (MPs) and nanoplastics (NPs): Introduction. In: Pandey Ashok, Tyagi R. D, Pandey Ashok, Drogui Patrick, Yadav Bhoomika, Pilli Sridhar, editors. Current Developments in Biotechnology and Bioengineering. Elsevier;2023. p. 1–32. DOI: 10.1016/B978-0-323-99908-3.00013-0. [Accessed: 2023-September-9]

[3] Kumar M, Chen H, Sarsaiya S, Qin S, Liu H, Awasthi MK, Kumar S, Singh L, Zhang Z, Bolan NS, Pandey A. Current research trends on micro-and nano-plastics as an emerging threat to global environment: A review. Journal of Hazardous Materials. 2021;**409**:124967. DOI: 10.1016/j.jhazmat.2020.124967

[4] Chícharo A, Barbosa AB. Hydrology and biota interactions as driving forces for ecosystem functioning. Treatise on Estuarine and Coastal Science. 2011;**10**:7–47. Available from: https://hdl.handle.net/10400.1/1904

[5] Dayal L, Raj D, Kumari P, Sinha S. Abundance of microplastics in marine and freshwater ecosystem and it's impact on biotic and abiotic components. Water, Air, and Soil Pollution. 2024;**235**(7):416. DOI: 10.1007/s11270-024-07199-x

[6] Pelegrini K, Pereira TCB, Maraschin TG, Teodoro LDS, Basso NRDS, De Galland GLB, Ligabue RA, Bogo MR. Micro-and nanoplastic toxicity: A review on size, type, source, and test-organism implications. Science of the Total Environment. 2023;**878**:162954. DOI: 10.1016/j.scitotenv.2023.162954

[7] Karnwal A, Martolia S, Dohroo A, Al-Tawaha ARMS, Malik T. Exploring bioremediation strategies for heavy metals and POPs pollution: The role of microbes, plants, and nanotechnology. Frontiers in Environmental Science. 2024;**12**:1397850. DOI: 10.3389/fenvs.2024.1397850

[8] Okoye CO, Addey CI, Oderinde O, Okoro JO, Uwamungu JY, Ikechukwu CK, Okeke ES, Ejeromedoghene O, Odii EC. Toxic chemicals and persistent organic pollutants associated with micro-and nanoplastics pollution. Chemical Engineering Journal Advances. 2022;**11**:100310. DOI: 10.1016/j.ceja.2022.100310

[9] Huang D, Chen H, Shen M, Tao J, Chen S, Yin L, Zhou W, Wang X, Xiao R, Li R. Recent advances on the transport of microplastics/nanoplastics in abiotic and biotic compartments. Journal of Hazardous Materials. 2022;**438**:129515. DOI: 10.1016/j.jhazmat.2022.129515

[10] Kumar M, Mrunalini K, Zhou Z, Modi RU, Bolan S, Rao CS, Braun M, Wrigley O, Li Y, Wang L, Leri A. Sources, distribution and migration of micro-nano-plastics from terrestrial to aquatic environments and its environmental and health impacts: A systematic review. Journal of Soils and Sediments. 2025;1–35. DOI: 10.1007/s11368-025-04042-6

[11] Ma C, Chen Q, Li J, Li B, Liang W, Su L, Shi H. Distribution and translocation of micro-and nanoplastics in fish. Critical Reviews in Toxicology. 2021;**51**(9):740–753. DOI: 10.1080/10408444.2021.2024495

[12] Kaur M, Sharma A, Bhatnagar P. Vertebrate response to microplastics, nanoplastics and co-exposed contaminants: Assessing accumulation, toxicity, behaviour, physiology, and molecular changes. Toxicology Letters. 2024. DOI: 10.1016/j.toxlet.2024.04.004

[13] Das A. The emerging role of microplastics in systemic toxicity: Involvement of reactive oxygen species (ROS). Science of the Total Environment. 2023;**895**:165076. DOI: 10.1016/j.scitotenv.2023.165076

[14] Sun R, Liu M, Xiong F, Xu K, Huang J, Liu J, Wang D, Pu Y. Polystyrene micro-and nanoplastics induce gastric toxicity through ROS mediated oxidative stress and P62/Keap1/Nrf2 pathway. Science of the Total Environment. 2024;**912**:169228. DOI: 10.1016/j.scitotenv.2023.169228

[15] Wei YY, Chen TT, Zhang DW, Zhang Y, Li F, Ding YC, Wang MY, Zhang L, Chen KG, Fei GH. Microplastics exacerbate ferroptosis via mitochondrial reactive oxygen species-mediated autophagy in chronic obstructive pulmonary disease. Autophagy. 2025;1–27. DOI: 10.1080/15548627.2025.2481126

[16] Hasan AM, Hamed M, Hasan J, Martyniuk CJ, Niyogi S, Chivers DP. A review of the neurobehavioural, physiological, and reproductive toxicity of microplastics in fishes. Ecotoxicology and Environmental Safety. 2024;**282**:116712. DOI: 10.1016/j.ecoenv.2024.116712

[17] Deidda I, Russo R, Bonaventura R, Costa C, Zito F, Lampiasi N. Neurotoxicity in marine invertebrates: An update. Biology. 2021;**10**(2):161. DOI: 10.3390/biology10020161

[18] Bhat RAH, Sidiq MJ, Altinok I. Impact of microplastics and nanoplastics on fish health and reproduction. Aquaculture. 2024;**741037**. DOI: 10.1016/j.aquaculture.2024.741037

[19] Zhang Q, Xu EG, Li J, Chen Q, Ma L, Zeng EY, Shi H. A review of microplastics in table salt, drinking water, and air: Direct human exposure. Environmental Science & Technology. 2020;**54**(7):3740–3751. DOI: 10.1021/acs.est.9b04535

[20] National Academies of Sciences, Engineering, and Medicine. Exposure to Contaminants Associated with Consumption of Seafood. In The Role of Seafood Consumption in Child Growth and Development. National Academies Press (US); 2024. DOI: 10.17226/27623

[21] Amaral-Zettler LA, Zettler ER, Mincer TJ. Ecology of the plastisphere. Nature Reviews Microbiology. 2020;**18**(3):139–151. DOI: 10.1038/s41579-019-0308-0

[22] Dey S, Rout AK, Behera BK, Ghosh K. Plastisphere community assemblage of aquatic environment: Plastic-microbe interaction, role in degradation and characterization technologies. Environmental Microbiome. 2022;**17**(1):32. DOI: 10.1186/s40793-022-00430-4

[23] Kanojia H, Purohit H, Joshi M, Kamdar JH, Chakraborty J. Microplastic to accumulate microbial pathogens in the terrestrial environment. In Microplastic

Pollution. Singapore: Springer Nature Singapore, 2024. p. 351–362. DOI: 10.1007/978-981-99-8357-5_20

[24] Jia J, Liu Q, Zhao E, Li X, Xiong X, Wu C. Biofilm formation on microplastics and interactions with antibiotics, antibiotic resistance genes and pathogens in aquatic environment. Eco-Environment & Health. 2024. DOI: 10.1016/j.eehl.2024.05.003

[25] Cholewińska P, Moniuszko H, Wojnarowski K, Pokorny P, Szeligowska N, Dobicki W, Polechoński R, Górniak W. The occurrence of microplastics and the formation of biofilms by pathogenic and opportunistic bacteria as threats in aquaculture. International Journal of Environmental Research and Public Health. 2022;**19**(13):8137. DOI: 10.3390/ijerph19138137

[26] Ferheen I, Spurio R, Marcheggiani S. Emerging Issues on Antibiotic-Resistant Bacteria Colonizing Plastic Waste in Aquatic Ecosystems. Antibiotics. 2024;**13**(4):339. DOI: 10.3390/antibiotics13040339

[27] Liu P, Zhan X, Wu X, Li J, Wang H, Gao S. Effect of weathering on environmental behavior of microplastics: Properties, sorption and potential risks. Chemosphere. 2020;**242**:125193. DOI: 10.1016/j.chemosphere.2019.125193

[28] Bhattacharya S, Kolandhasamy P, Mandal A, Rajaram R, Darbha GK. Ecological risk assessment and ingestion of microplastics in edible finfish and shellfish species collected from tropical mangrove forest, Southeastern India. Chemosphere. 2025;**377**:144308. DOI: 10.1016/j.chemosphere.2025.144308

[29] Wang L, Wu WM, Bolan NS, Tsang DC, Li Y, Qin M, Hou D. Environmental fate, toxicity and risk management strategies of nanoplastics in the environment: Current status and future perspectives. Journal of Hazardous Materials. 2021;**401**:123415. DOI: 10.1016/j.jhazmat.2020.123415

[30] Kampouris TE, Syranidou E, Seridou P, Gagoulis K, Batjakas IE, Kalogerakis N. MPs and NPs intake and heavy metals accumulation in tissues of Palinurus elephas (JC Fabricius, 1787), from NW Aegean sea, Greece. Environmental Pollution. 2023;**316**:120725. DOI: 10.1016/j.envpol.2022.120725

[31] Kumar PS, Korving L, Keesman KJ, van Loosdrecht MC, Witkamp GJ. Effect of pore size distribution and particle size of porous metal oxides on phosphate adsorption capacity and kinetics. Chemical Engineering Journal. 2019;**358**:160–169. DOI: 10.1016/j.cej.2018.09.202

[32] Trestrail C, Walpitagama M, Hedges C, Truskewycz A, Miranda A, Wlodkowic D, Shimeta J, Nugegoda D. Foaming at the mouth: Ingestion of floral foam microplastics by aquatic animals. Science of the Total Environment. 2020;**705**:135826. DOI: 10.1016/j.scitotenv.2019.135826

[33] Merino-Guzmán C, Jofré-Fernández I. Polymer-specific impacts of microplastics on mineral retention and soil stability. Environmental Pollution. 2025;**126487**. DOI: 10.1016/j.envpol.2025.126487

[34] Veloso RCDS, 2024. Study of the Influence of Fluvial Dynamics on the Distribution and Transport of Microplastics (Doctoral dissertation). Available from: https://hdl.handle.net/

1822/92334 [Accessed: 2024-January-17]

[35] Ermis H, Collins C, Saha SK, Murray P. Beyond Visibility: Microorganisms for tackling plastic and microplastic problems for cleaner future. Chemical Engineering Journal. 2024:154585. DOI: 10.1016/j.cej.2024.154585

[36] Liang J, Chen X, Duan X, Gu X, Zhao X, Zha S, Chen X. Natural aging and adsorption/desorption behaviors of polyethylene mulch films: Roles of film types and exposure patterns. Journal of Hazardous Materials. 2024;**466**:133588. DOI: 10.1016/j.jhazmat.2024.133588

[37] Khaki QZ, Kumar P. Ecological impacts of microplastics and their additives: Exposure risk/toxicity assessment and fate/transport of persistent, bio-accumulative and toxic substances. Microplastics in the Environment: Fate, Impacts, Removal, and Management. 2025;259–282. DOI: 10.1002/9781394251100.ch10

[38] Ji G, Xing Y, You T. Biochar as adsorbents for environmental microplastics and nanoplastics removal. Journal of Environmental Chemical Engineering. 2024:113377. DOI: 10.1016/j.jece.2024.113377

[39] Rayhan MRI, Shohag AS, Riya KA, Liza JM, Rahman MM, Rahaman MS. Legacy and emerging contaminants. Discussions and Legislative Advances. 2024. DOI: 10.1007/698_2024_1176

[40] Wong MH. Managing Legacy POPs and Emerging Chemicals: Insights from UNEP/GEF Projects. In Ecological and Human Health Impacts of Contaminated Food and Environments 1st Edition , Boca Raton: eBook, 2025. p. 1. DOI: 10.1201/9781003492115

[41] da Luz TM. Ecotoxicological effects of polycyclic aromatic hydrocarbons in aquatic organisms. 2025. DOI: 10.1007/698_2024_1181

[42] Zhang BF, Liu Y, Pei DS. Principal sources, fate, and mechanism of chemical mixtures in the environment. Toxicological Assessment of Combined Chemicals in the Environment. 2025;47–60. DOI: 10.1002/9781394158355.ch04

[43] Wang Y, Nan X, Sun H, Shi Y, Miao J, Li Y, Han X, Zhang N, Wang H, Ren N, Zhao X. From insects to mammals! Tissue accumulation and transgenerational transfer of micro/nano-plastics through the food chain. Journal of Hazardous Materials. 2024;**480**:136424. DOI: 10.1016/j.jhazmat.2024.136424

[44] Bhaumik S, Chakraborty P. Interactions between microplastics (MPs) and trace/toxic metals in marine environments: Implications and insights —A comprehensive review. Environmental Science and Pollution Research. 2024;**31**(50):59681–59699. DOI: 10.1007/s11356-024-34960-w

[45] Dong Z, Wang WX. Modeling the vertical transport of copepod fecal particles under nano/microplastic exposure. Environmental Science & Technology. 2025;**59**(13):6610–6622. DOI: 10.1021/acs.est.5c01967

[46] Sandil S, Zaray G. Impact of microplastics and nanoplastics in the aquatic environment. Management of Micro and Nano-plastics in Soil and Biosolids: Fate, Occurrence, Monitoring, and Remedies. 2024;25–68. DOI: 10.1007/978-3-031-51967-3_2

[47] Balog S, de Almeida MS, Taladriz-Blanco P, Rothen-Rutishauser B, Petri-

Fink A. Does the surface charge of the nanoparticles drive nanoparticle–cell membrane interactions? Current Opinion in Biotechnology. 2024;**87**:103128. DOI: 10.1016/j.copbio.2024.103128

[48] Lee H, Song SJ, Kim CS, Park B. Polystyrene nanoplastics-induced intestinal barrier disruption via inflammation and apoptosis in zebrafish larvae (Danio Rerio). Aquatic Toxicology. 2024;**274**:107027. DOI: 10.1016/j.aquatox.2024.107027

[49] Jamil H, Kamal S, Lodhi MS, Wajee W, Nidaullah H, Kabir A, Naseem F, Mushtaq B, Ashraf N. Impacts of polystyrene nanoplastics on fisheries biology and prospective remediation approaches in aquatic ecosystems. Indus Journal of Bioscience Research. 2025;**3**(5):75–86. DOI: 10.70749/ijbr.v3i5.1491.

[50] Junaid M, Hamid N, Liu S, Abbas Z, Imran M, Haider MR, Wang B, Chen G, Khan HK, Yue Q, Xu N. Interactive impacts of photoaged micro (nano) plastics and co-occurring chemicals in the environment. Science of the Total Environment. 2024;**172213**. DOI: 10.1016/j.scitotenv.2024.172213

[51] Zhang X, Yin Z, Xiang S, Yan H, Tian H. Degradation of polymer materials in the environment and its impact on the health of experimental animals: A review. Polymers. 2024;**16**(19):2807. DOI: 10.3390/polym16192807

[52] Habumugisha T, Zhang Z, Uwizewe C, Yan C, Ndayishimiye JC, Rehman A, Zhang X. Toxicological review of micro-and nano-plastics in aquatic environments: Risks to ecosystems, food web dynamics and human health. Ecotoxicology and Environmental Safety. 2024;**278**:116426. DOI: 10.1016/j.ecoenv.2024.116426

[53] Wu Q, Cao J, Liu X, Zhu X, Huang C, Wang X, Song Y. Micro (nano)-plastics exposure induced programmed cell death and corresponding influence factors. Science of the Total Environment. 2024;**171230**. DOI: 10.1016/j.scitotenv.2024.17123

[54] Mahmud F, Sarker DB, Jocelyn JA, Sang QXA. Molecular and cellular effects of microplastics and nanoplastics: Focus on inflammation and senescence. Cells. 2024;**13**(21):1788. DOI: 10.3390/cells13211788

[55] Fan J, Liu L, Lu Y, Chen Q, Fan S, Yang Y, Long Y, Liu X. Acute exposure to polystyrene nanoparticles promotes liver injury by inducing mitochondrial ROS-dependent necroptosis and augmenting macrophage-hepatocyte crosstalk. Particle and Fibre Toxicology. 2024;**21**(1):20

[56] Wang YL, Lee YH, Chou CL, Chang YS, Liu WC, Chiu HW. Oxidative stress and potential effects of metal nanoparticles: A review of biocompatibility and toxicity concerns. Environmental Pollution. 2024:123617. DOI: 10.1016/j.envpol.2024.123617

[57] Uti DE, Atangwho IJ, Alum EU, Ntaobeten E, Obeten UN, Bawa I, Agada SA, Ukam CIO, Egbung GE. Antioxidants in cancer therapy mitigating lipid peroxidation without compromising treatment through nanotechnology. Discover Nano. 2025;**20**(1):1–43. DOI: 10.1186/s11671-025-04248-0

[58] Sales Junior SF, da Costa NM, de Farias Araújo G, Soares LOS, Mannarino CF, Correia FV,

Saggioro EM. Antioxidant system alterations, oxidative, and genotoxic effects in Danio rerio (zebrafish) exposed to leachate from a dumpsite. Environmental Science and Pollution Research. 2024;**31**(7):10737–10749. DOI: 10.1007/s11356-024-31883-4

[59] Copeto S, Ganço S, Ferreira IJ, Silva M, Motta C, Diniz M. The effects of tetrabromobisphenol A (TBBPA) on the mussel mytilus galloprovincialis: A multi-biomarker approach. Oceans. 2024;**5**(2):181–195. MDPI. DOI: 10.3390/oceans5020011

[60] Copeto S, Ganço S, Ferreira IJ, Sanchez D, Nunes MJ, Motta C, Silva M, Diniz M. The impact of perfluorooctanoic acid (PFOA) on the mussel mytilus galloprovincialis: A multi-biomarker evaluation. Oceans. 2024;**5**(4):857–873. DOI: 10.3390/oceans5040049

[61] Summer M, Ashraf R, Ali S, Bach H, Noor S, Noor Q, Riaz S, Khan RRM. Inflammatory response of nanoparticles: Mechanisms, consequences, and strategies for mitigation. Chemosphere. 2024:142826. DOI: 10.1016/j.chemosphere.2024.142826

[62] Shrivastava Y, Nayak SK, Mogalekar HS. Aquatic Animal Health Management. In Management of Fish Diseases. Singapore: Springer Nature Singapore, 2025. p. 189–211. DOI: 10.1007/978-981-96-0270-4_8

[63] Rodríguez-González J, Gutiérrez-Kobeh L. Apoptosis and its pathways as targets for intracellular pathogens to persist in cells. Parasitology Research. 2024;**123**(1):60. DOI: 10.1007/s00436-023-08031-x

[64] Zapata-Lopera YM, Trejo-Tapia G, Cano-Europa E, Rodríguez-Hernández AA, Rojas-Franco P, Herrera-Ruiz M, Jiménez-Ferrer E. Neuroprotective effect of Bouvardia ternifolia (Cav.) Schltdl via inhibition of TLR4/NF-κB, caspase-3/Bax/Bcl-2 pathways in ischemia/reperfusion injury in rats. Frontiers in Pharmacology. 2024;**15**:1471542. DOI: 10.3389/fphar.2024.1471542

[65] Tang KHD. Genotoxicity of microplastics on living organisms: effects on chromosomes, DNA and gene expression. Environments. 2025;**12**(1):10. DOI: 10.3390/environments12010010

[66] Borea II, 2024. Impacts of Petroleum Exposure on DNA Integrity and Gene Expression in the Pacific Oyster, Crassostrea Gigas. Thesis. (Project) M.E.T. 2024-02-28. etd22930

[67] Spurgeon D, Track 1. environmental and human toxicology: From molecules to organisms, from omics to in Vivo. fully dynamic carbon footprint of circular biobased systems– A framework with temporal life cycle inventory database (DyPLCA) tailored for forestry and wood products cascades, p.5. PRINT ISSN 2309-8031 - ONLINE ISSN 2310-3043

[68] Ali W, Buriro RS, Gandahi JA, Chen Y, Ul Aabdin Z, Bhutto S, Sun J, Zhu J, Liu Z, Zou H. A critical review on male-female reproductive and developmental toxicity induced by micro-plastics and nano-plastics through different signaling pathways. Chemico-biological Interactions. 2024:110976. DOI: 10.1016/j.cbi.2024.110976

[69] Madesh S, Sudhakaran G, Murugan R, Almutairi MH, Almutairi BO, Kathiravan MK, Arockiaraj J. Parental (F0) exposure to

Cadmium and Ketoprofen induces developmental deformities in offspring (F1): A transgenerational toxicity assessment in zebrafish model. Science of the Total Environment. 2024;**950**:175319. DOI: 10.1016/j.scitotenv.2024.175319

[70] Guerrero Limon G, 2024. Evaluating the e. ects of mixtures of endocrine disruptors on development and gene expression in zebrafish larvae (Danio rerio). Available from: https://hdl.handle.net/2268/325652 [Accessed: 2024-December-20]

[71] Ye R, Li Z, Xian H, Zhong Y, Liang B, Huang Y, Chen D, Dai M, Tang S, Guo J, Bai R. Combined effects of polystyrene nanosphere and homosalate exposures on estrogenic end points in MCF-7 cells and zebrafish. Environmental Health Perspectives. 2024;**132**(2):027011. DOI: 10.1289/EHP13696

[72] Ojo AB, Agbeye OD, Ogwa TO, Adedoyin D, Rotimi DE, Ojo OA. Implications of plastic-derived endocrine disruptors on human health. Toxicology Mechanisms and Methods. 2025;1–25. DOI: 10.1080/15376516.2025.2510525

[73] Liu S, He Y, Yin J, Zhu Q, Liao C, Jiang G. Neurotoxicities induced by micro/nanoplastics: A review focusing on the risks of neurological diseases. Journal of Hazardous Materials. 2024:134054. DOI: 10.1016/j.jhazmat.2024.134054

[74] Porras-Rivera G, Górski K, Colin N. Behavioral biomarkers in fishes: A non-lethal approach to assess the effects of chemical pollution on freshwater ecosystems. Environmental Research. 2024;**119607**. DOI: 10.1016/j.envres.2024.119607

[75] Esterhuizen M, Lee SA, Kim Y, Järvinen R, Kim YJ. Ecotoxicological consequences of polystyrene naturally leached in pure, fresh, and saltwater: Lethal and nonlethal toxicological responses in Daphnia magna and Artemia salina. Frontiers in Marine Science. 2024;**11**:1338872. DOI: 10.3389/fmars.2024.1338872

[76] D'Souza LC, Paithankar JG, Stopper H, Pandey A, Sharma A. Environmental chemical-induced reactive oxygen species generation and immunotoxicity: A comprehensive review. Antioxidants and Redox Signaling. 2024;**40**(10-12):691–714. DOI: 10.1089/ars.2022.0117

[77] Song Q, Li Q, Yang Y, Gao H, Han F. Antimicrobial functions of galectins from fish, mollusks, and crustaceans: A review. Journal of Agricultural and Food Chemistry. 2024;**72**(45):24895–24907. DOI: 10.1021/acs.jafc.4c05412

[78] Guo Z, Li J, Zhang Z. Meta-analysis for systematic review of global micro/nano-plastics contamination versus various freshwater microalgae: Toxicological effect patterns, taxon-specific response, and potential eco-risks. Water Research. 2024:121706. DOI: 10.1016/j.watres.2024.121706

[79] Zhao B, Chen F, Yao Q, Lin M, Zhou K, Mi S, Pan H, Zhao X. Toxicity effects and mechanism of micro/nanoplastics and loaded conventional pollutants on zooplankton: An overview. Marine Environmental Research. 2024;**106547**. DOI: 10.1016/j.marenvres.2024.106547

[80] Zhang B, Wang Z, Li D, Li L, Zhao Y, Tang X, Zhao Y. Reactive oxygen species mediated extracellular

polymeric substances production assisting the recovery of Thalassiosira pseudonana from polystyrene micro and nanoplastics exposure. Environmental Pollution. 2024;**348**:123850. DOI: 10.1016/j.envpol.2024.123850

[81] Yang W, Gao P, Ye Z, Chen F, Zhu L. Micro/nano-plastics and microalgae in aquatic environment: Influence factor, interaction, and molecular mechanisms. Science of the Total Environment. 2024;**173218**. DOI: 10.1016/j.scitotenv.2024.173218

[82] Zheng H, Li Q, Gu X, Qian Y, Shang Y, Yu H, An L, Wang F. Toxicity and health effects of microplastics. Microplastics in the Environment: From Formation to Remediation. 2025;137–244. DOI: 10.1002/9783527842124.ch3

[83] de Ruijter VN, 2024. Improving the Realism and the Relevance of Microplastic Effect Tests (Doctoral Dissertation, Wageningen University and Research). DOI: 10.18147/656569

[84] Vagi MC, Petsas AS, Kostopoulou MN. Potential effects of persistent organic contaminants on marine biota: A review on recent research. Water. 2021;**13**(18):2488. DOI: 10.3390/w13182488

[85] Amiard JC, Chaumot A, Couderc M, Garric J, Geffard O, Xuereb B, 2012. Impairments of endocrine functions: Causes and consequences. 32.9780429111495

[86] Kakakhel MA, Narwal N, Kataria N, Johari SA, Din SZU, Jiang Z, Khoo KS, Xiaotao S. Deciphering the dysbiosis caused in the fish microbiota by emerging contaminants and its mitigation strategies-a review. Environmental Research. 2023; **237**:117002. DOI: 10.1016/j.envres.2023.117002

[87] Amereh F, Babaei M, Eslami A, Fazelipour S, Rafiee M. The emerging risk of exposure to nano (micro) plastics on endocrine disturbance and reproductive toxicity: From a hypothetical scenario to a global public health challenge. Environmental Pollution. 2020;**261**:114158. DOI: 10.1016/j.envpol.2020.114158

[88] He Y, Yin R. The reproductive and transgenerational toxicity of microplastics and nanoplastics: A threat to mammalian fertility in both sexes. Journal of Applied Toxicology. 2024;**44** (1):66–85. DOI: 10.1002/jat.4510

[89] Saha G, Chandrasekaran N. A comprehensive review on the adverse effect of microplastics in the gastrointestinal system of Artemia sp. Heliyon. 2024. DOI: 10.1016/j.heliyon.2024.e37720

[90] Banaee M, Multisanti CR, Impellitteri F, Piccione G, Faggio C. Environmental toxicology of microplastic particles on fish: A review. Comparative Biochemistry and Physiology Part C: Toxicology & Pharmacology. 2024;**110042**. DOI: 10.1016/j.cbpc.2024.110042

[91] Soliño L, Turner AD, Ben-Gigirey B, Alexander RP, Dean KJ, Hatfield RG, Maskrey BH, Casero MVM. Investigation into paralytic shellfish toxins and microcystins in seabirds from Portugal. Toxins. 2025;**17**(3):135. DOI: 10.3390/toxins17030135

[92] Villamarín F, Jardine TD, Bunn SE, Malvasio A, Piña CI, Jacobi CM, Araújo DD, de Brito ES, de Moraes Carvalho F, da Costa ID, Verdade LM. Body size predicts

ontogenetic nitrogen stable-isotope (δ 15N) variation, but has little relationship with trophic level in ectotherm vertebrate predators. Scientific Reports. 2024;**14**(1):14102. DOI: 10.1038/s41598-024-61969-5

[93] Li X, Dai Y, Li X, Guo H, Dai J, Wang H, Xiong D, Liao G. Oxidative stress responses and recovery of marine medaka (Oryzias melastigma) in early-life stages after acute exposure to crude oil. Journal of Marine Science and Engineering. 2025;**13**(5):965. DOI: 10.3390/jmse13050965

[94] Oros A. Bioaccumulation and trophic transfer of heavy metals in marine fish: Ecological and ecosystem-level impacts. Journal of Xenobiotics. 2025;**15**(2):59. DOI: 10.3390/jox15020059

[95] Alkhair S, Zouari N, Ibrahim MIA, Al-Ghouti MA. Efficacy of adsorption processes employing green nanoparticles for bisphenol A decontamination in water: A review. Environmental Nanotechnology, Monitoring & Management. 2024:100963. DOI: 10.1016/j.enmm.2024.100963

[96] Marcellus KA, Prescott D, Scur M, Ross N, Gill SS. Exposure of polystyrene nano-and microplastics in increasingly complex in vitro intestinal cell models. Nanomaterials. 2025;**15**(4):267. DOI: 10.3390/nano15040267

[97] Raymond JA, 2024. Tracking Plastic Particles and Tributyltin in Marine Invertebrates (Doctoral Dissertation, University of Surrey).

[98] Rosner A, Ballarin L, Barnay-Verdier S, Borisenko I, Drago L, Drobne D, Concetta Eliso M, Harbuzov Z, Grimaldi A, Guy-Haim T, Karahan A. A broad-taxa approach as an important concept in ecotoxicological studies and pollution monitoring. Biological Reviews. 2024;**99**(1):131–176. DOI: 10.1111/brv.13015

[99] Peacock MM. Negotiating a fragmented world: What do we know, how do we know it, and where do we go from here? Diversity. 2025;**17**(3):200. DOI: 10.3390/d17030200

[100] Das BK, Santra N, and Samanta G. Exploring dynamics of predator-prey interactions. In Filomat. University of Nis, Serbia: Faculty of Sciences and Mathematics. **38**; 2024. p. 11061–11083 DOI: 10.2298/FIL2431061D

[101] Yadav DK, Samantaray BP, Kumar R. Effect of alternative natural diet on microplastic ingestion, functional responses and trophic transfer in a tri-trophic coastal pelagic food web. Science of the Total Environment. 2024;**950**:174999. DOI: 10.1016/j.scitotenv.2024.174999

[102] Ahmed R, Tamim MTR. Marine and coastal environments: Challenges, impacts, and strategies for a sustainable future. International Journal of Science Education and Science. 2025;**2**(1):53–60. DOI: 10.56566/ijses.v2i1.325

[103] Li X, Shen X, Jiang W, Xi Y, Li S. Comprehensive review of emerging contaminants: Detection technologies, environmental impact, and management strategies. Ecotoxicology and Environmental Safety. 2024;**278**:116420. DOI: 10.1016/j.ecoenv.2024.116420

[104] Barhoumi B, Metian M, Alshemmari H, Chouba L, Al-Rawabdeh AM, Touil S, Alonso-Hernández CM, Driss MR, Bersuder P, Microplastics in tunisian aquatic ecosystems: Current research trends, knowledge gaps, and

prospects. Handbook of Microplastic Pollution in the Environment, pp.590–630.

[105] Jeong E, Lee JY, Redwan M. Animal exposure to microplastics and health effects: A review. Emerging Contaminants. 2024:100369. DOI: 10.1016/j.emcon.2024.100369

[106] Mortuza A, Meese EN, Wharton M, Gahn B, Petersen L, Quigg A, Wells RJD, Kaiser K, and Hala D. Comparing Persistent (Polycyclic Aromatic Hydrocarbons and Polychlorinated Biphenyls) and Emerging Nano(Micro)Plastics (Nmps) Pollutant Body-Burdens in Oysters and Fish from Matagorda Bay. Texas, USA: Marine Pollution Bulletin; 2025;**221**. DOI: 10.2139/ssrn.5229642 [Accessed: 2025-July-28]

[107] Zhang L, Wu C, Wang Q. Toxicity of engineered nanoparticles in food: sources, mechanisms, contributing factors, and assessment techniques. Journal of Agricultural and Food Chemistry. 2025. DOI: 10.1021/acs.jafc.5c01550

[108] Alijagic A, Suljević D, Fočak M, Sulejmanović J, Šehović E, Särndahl E, and Engwall M . The Triple Exposure Nexus of Microplastic Particles, Plastic-associated Chemicals, and Environmental Pollutants from a Human Health Perspective. Sweden, Bosnia and Herzegovina: Environment International; 2024. Vol. **188**. p. 108736 . DOI: 10.1016/j.envint.2024.108736

[109] Samaei SHA, Mojahednia P, Chen J, Li Z, Jaszczyszyn K, Kiedrzyńska E, Xue J. What does the "Trojan Horse" Carry? the pollutants associated with microplastics/nanoplastics in water environments. ACS Environmental Science & Technology Water. 2025;**5**(4):1530–1545. DOI: 10.1021/acsestwater.4c01089

[110] Taylor B, Ofori KF, Parsaeimehr A, Akdemir Evrendilek G, Attarwala T, Ozbay G. Exploring the complexities of seafood: From benefits to contaminants. Foods. 2025;**14**(9):1461. DOI: 10.3390/foods14091461

[111] Xu H, Yu Z, Dong C, Yao W, Xie Y. Cap opening induced microplastic contamination in bottled water revealed by single particle FTIR imaging and analysis. Food Control. 2025:111508. DOI: 10.1016/j.foodcont.2025.111508

[112] Tan H, Kek HY, Van Fan Y, Sheng DDCV, Lee CH, Tan KY, Rani MFH, Wong KY. Systematic review on indoor microplastics: Unveiling sources, exposure pathways, and human health implications. Journal of Advanced Research in Micro and Nano Engineering. 2024;**20**(1):96–111. DOI: 10.37934/armne.20.1.96111

[113] Chen S, 2024. A One Health perspective on bacterial zoonotic risks due to environmental contamination at public outdoor events. Available from: https://studenttheses.uu.nl/handle/20.500.12932/47580 [Accessed: 2024-August-31]

[114] Lyu Q, Nie Y, Gao J, Wang D. The association between urinary BPA concentrations and urinary incontinence in women. Scientific Reports. 2025;**15**(1):1–8. DOI: 10.1038/s41598-025-99079-5

[115] Kumar V, Sharma N, Umesh M, Gupta P, Sharma P, Basheer T, Huligowda LKD, Thomas J, Bhagat SK, Pasrija R. Microplastics in food: Occurrence, toxicity, green analytical detection methods and future

challenges. Green Analytical Chemistry. 2024;**11**:100152. DOI: 10.1016/j.greeac.2024.100152

[116] LeChevallier MW, Prosser T, Stevens M. Opportunistic pathogens in drinking water distribution systems—A review. Microorganisms. 2024;**12**(5):916. DOI: 10.3390/microorganisms12050916

[117] Khadanga V, Mishra PC. A review on toxicity mechanism and risk factors of nanoparticles in respiratory tract. Toxicology. 2024:153781. DOI: 10.1016/j.tox.2024.153781

[118] Kisielinski K, Hockertz S, Hirsch O, Korupp S, Klosterhalfen B, Schnepf A, Dyker G. Wearing face masks as a potential source for inhalation and oral uptake of inanimate toxins–A scoping review. Ecotoxicology and Environmental Safety. 2024;**275**:115858. DOI: 10.1016/j.ecoenv.2023.115858

[119] Kim H, Park HH, Kim HN, Seo D, Hong KS, Jang JG, Seo EU, Kim IY, Jeon SY, Son B, Cho SW. The TOX–RAGE axis mediates inflammatory activation and lung injury in severe pulmonary infectious diseases. Proceedings of the National Academy of Sciences. 2024;**121**(26):e2319322121. DOI: 10.1073/pnas.231932212

[120] Vashishat A, Patel P, Das Gupta G, Das Kurmi B. Alternatives of animal models for biomedical research: A comprehensive review of modern approaches. Stem Cell Reviews and Reports. 2024;**20**(4):881–899. DOI: 10.1007/s12015-024-10701-x

[121] Fujihara J, Nishimoto N. Review of zinc oxide nanoparticles: Toxicokinetics, tissue distribution for various exposure routes, toxicological effects, toxicity mechanism in mammals, and an approach for toxicity reduction. Biological Trace Element Research. 2024;**202**(1):9–23. DOI: 10.1007/s12011-023-03644-w

[122] Lin H, Li X, Gao H, Hu W, Yu S, Li X, Lei L, Yang F. The role of gut microbiota in mediating increased toxicity of nano-sized polystyrene compared to micro-sized polystyrene in mice. Chemosphere. 2024;**358**:142275. DOI: 10.1016/j.chemosphere.2024.142275

[123] Jin W, Zhang W, Tang H, Wang P, Zhang Y, Liu S, Qiu J, Chen H, Wang L, Wang R, Sun Y. Microplastics exposure causes the senescence of human lung epithelial cells and mouse lungs by inducing ROS signaling. Environment International. 2024;**185**:108489. DOI: 10.1016/j.envint.2024.108489

[124] Zhi L, Li Z, Su Z, Wang J. Immunotoxicity of microplastics: Carrying pathogens and destroying the immune system. TrAC Trends in Analytical Chemistry. 2024;**177**:117817. DOI: 10.1016/j.trac.2024.117817

[125] Bhol NK, Bhanjadeo MM, Singh AK, Dash UC, Ojha RR, Majhi S, Duttaroy AK, Jena AB. The interplay between cytokines, inflammation, and antioxidants: Mechanistic insights and therapeutic potentials of various antioxidants and anti-cytokine compounds. Biomedicine & Pharmacotherapy. 2024;**178**:117177. DOI: 10.1016/j.biopha.2024.117177

[126] Duan L, Hao Z, Ji R, Li X, Wang H, Su Y, Guan F, Ma S. Glucose-modified BSA/procyanidin C1 NPs penetrate the blood-brain barrier and alleviate neuroinflammation in Alzheimer's disease models.

International Journal of Biological Macromolecules. 2024;**268**:131739. DOI: 10.1016/j.ijbiomac.2024.131739

[127] Kaushik A, Singh A, Gupta VK, Mishra YK. Nano/micro-plastic, an invisible threat getting into the brain. Chemosphere. 2024;**361**:142380. DOI: 10.1016/j.chemosphere.2024.142380

[128] Mans RA, Kelehear H, Rotschafer S, Ganas C, Uche-Moon B, Call G, Mauersberg CC, Toller J, Diamanduros A. Exposure to a nanoplastic-enriched diet for fourteen days increases microglial immunoreactivity in the zebrafish telencephalon. Frontiers in Molecular Neuroscience. 2025;**18**:1563086. DOI: 10.3389/fnmol.2025.1563086

[129] Kelehear HA, "Evaluating reactive gliosis in the brain of adult zebrafish following the ingestion of nanopolystyrene." (2024). Electronic Theses and Dissertations. 2853. Society for Neuroscience. DOI: 10.1152/physiol.2024.39.S1.210. Available from: https://digitalcommons.georgiasouthern.edu/etd/2853 [Accessed: 2024-April-7]

[130] Chae YR, Lee YR, Kim YS, Park HY. Diet-induced gut dysbiosis and leaky gut syndrome. Journal of Microbiology and Biotechnology. 2024;**34**(4):747. DOI: 10.4014/jmb.2312.12031

[131] Haney J, Rochman CM. Plastic pollution has the potential to alter ecological and evolutionary processes in aquatic ecosystems. Nature Ecology & Evolution. 2025;1–7. DOI: 10.1016/j.greeac.2024.100152

[132] Zhao Z, Wang Y, Wei Y, Peng G, Wei T, He J, Li R, Wang Y. Distinctive patterns of bacterial community succession in the riverine micro-plastisphere in view of biofilm development and ecological niches. Journal of Hazardous Materials. 2024;**480**:135974. DOI: 10.1016/j.jhazmat.2024.135974

[133] Mishra S, Dash D, Das AP. Aquatic microbial diversity on plastisphere: Colonization and potential role in microplastic biodegradation. Geomicrobiology Journal. 2024;**41**(4):312–323. DOI: 10.1080/01490451.2023.2209750

[134] Saini P, Mishra P. Biofilm linked microbial prospecting of bioremediation. In Bioprospecting of Microbial Resources for Agriculture, Environment and Bio-chemical Industry. Cham: Springer Nature Switzerland, 2024. p. 87–108. DOI: 10.1007/978-3-031-63844-2_5

[135] Atmakuri A, Yadav B, Tiwari B, et al. Nature's architects: A comprehensive review of extracellular polymeric substances and their diverse applications. Waste Disposal and Sustainable Energy. 2024;**6**:529–551

[136] MacLean J, Mayanna S, Benning LG, Horn F, Bartholomäus A, Wiesner Y, Wagner D, Liebner S. The terrestrial plastisphere: Diversity and polymer-colonizing potential of plastic-associated microbial communities in soil. Microorganisms. 2021;**9**(9):1876. DOI: 10.3390/microorganisms9091876

[137] Yu S, Lu X, Lu H. Marine microbial biofilms on diverse abiotic surfaces. Frontiers in Marine Science. 2025;**12**:1482946. DOI: 10.3389/fmars.2025.1482946

[138] Shafana Farveen M, Narayanan R. Omic-driven strategies to unveil

microbiome potential for biodegradation of plastics: A review. Archives of Microbiology. 2024;**206**(11):441. DOI: 10.1007/s00203-024-04165-3

[139] Macey MC. Genome-resolved metagenomics identifies novel active microbes in biogeochemical cycling within methanol-enriched soil. Environmental Microbiology Reports. 2024;**16**(2):e13246. DOI: 10.1111/1758-2229.13246

[140] Nene A, Sadeghzade S, Viaroli S, Yang W, Uchenna UP, Kandwal A, Liu X, Somani P, Galluzzi M. Recent advances and future technologies in nano-microplastics detection. Environmental Sciences Europe. 2025;**37**(1):7. DOI: 10.1186/s12302-024-01044-y

[141] Barry PJ, Curd A, Droual G, le Moigne M, Plastisphere nomads: An overview of organism dispersal via marine plastic litter. Plastisphere, 328–351

[142] Yan X, Chio C, Li H, Zhu Y, Chen X, Qin W. Colonization characteristics and surface effects of microplastic biofilms: Implications for environmental behavior of typical pollutants. Science of the Total Environment. 2024:173141. DOI: 10.1016/j.scitotenv.2024.173141

[143] Mulky L, Valiaparampil JG, Somashekara D. Plastisphere-a new habitat of microbial community: Composition, structure and ecological consequences. Sustainable Environment. 2025;**11**(1):2516898. DOI: 10.1080/27658511.2025.2516898

[144] Wang Y, Liu X, Huang C, Han W, Gu P, Jing R, Yang Q. Antibiotic resistance genes and virulence factors in the plastisphere in wastewater treatment plant effluent: Health risk quantification and driving mechanism interpretation. Water Research. 2025;**271**:122896. DOI: 10.1016/j.watres.2024.122896

[145] Ferheen I, Spurio R, Marcheggiani S. Vehicle transmission of antibiotic-resistant pathogens mediated by plastic debris in aquatic ecosystems. Iscience. 2024;**27**(6). DOI: 10.1016/j.isci.2024.110026

[146] Magana-Arachchi DN, Bandara T, and Gunathilaka S. Plastisphere as a Vector for Pathogenic Microbes and Antibiotic Resistance. In Plastisphere 1st Edition, Boca Raton: CRC Press. eBook, 2025. p. 312–327. DOI: 10.1201/9781032717814

[147] Cholewińska P, Wojnarowski K, Szeligowska N, Pokorny P, Hussein W, Hasegawa Y, Dobicki W, Palić D. Presence of microplastic particles increased abundance of pathogens and antimicrobial resistance genes in microbial communities from the Oder river water and sediment. Scientific Reports. 2025;**15**(1):1–19. DOI: 10.1038/s41598-025-01136-6

[148] Ormsby MJ, Woodford L, White HL, Fellows R, Quilliam RS. The plastisphere can protect salmonella typhimurium from UV stress under simulated environmental conditions. Environmental Pollution. 2024;**358**:124464. DOI: 10.1016/j.envpol.2024.124464

[149] Lu H, Wang Y, Liu H, Wang N, Zhang Y, Li X. Review of the presence and phage-mediated transfer of ARGs in biofilms. Microorganisms. 2025;**13**(5):997

[150] Armanni A, Fumagalli S, Ghisleni G, Corneo L, Zenaro M, Casiraghi M, Labra M, Colombo M, Bruno A, 2024. Characterization of

hospital and patients' microbiome biodiversity for nano-detectors development and antibiotic resistance monitoring. Available from: https://hdl.handle.net/10281/485199

[151] Motlhalamme T, Paul L, Singh V. Environmental Reservoirs, Genomic Epidemiology, and Mobile Genetic Elements. In Antimicrobial Resistance: Factors to Findings: Omics and Systems Biology Approaches. Cham: Springer International Publishing, 2024. p. 239–273. DOI: 10.1007/978-3-031-65986-7_7

[152] Chibwe M, Odume ON, Nnadozie CF. Campylobacter—an emerging pollutant of aquatic environments. Emerging Pollutants: Protecting Water Quality for the Health of People and the Environment. 2025;35–65. DOI: 10.1007/978-3-031-71758-1

[153] Rathod SV, Saras P, and Gondaliya SM. . doi: . Environmental pollution: Threats and challenges for management Eco-Restoration of Polluted Environment. CRC Press. eBook. 2024:1–34. DOI: 10.1201/9781003423393

[154] Rodrigues PM, Pinto FS, Marques RC. A framework for enabling conditions for wastewater reuse. Sustainable Production and Consumption. 2024;**46**:355–366. DOI: 10.1016/j.spc.2024.02.022

[155] Jahedi F, Fard NJH. Micro-and nanoplastic toxicity in humans: Exposure pathways, cellular effects, and mitigation strategies. Toxicology Reports. 2025:102043. DOI: 10.1016/j.toxrep.2025.102043

[156] Sinha RK, Kumar R, Phartyal SS, Sharma P. Interventions of citizen science for mitigation and management of plastic pollution: Understanding sustainable development goals, policies, and regulations. Science of the Total Environment. 2024:176621. DOI: 10.1016/j.scitotenv.2024.176621

[157] de Ruijter VN, Redondo-Hasselerharm PE, Koelmans AA. A brief history of microplastics effect testing: Guidance and prospect. Environmental Pollution. 2025:125711. DOI: 10.1016/j.envpol.2025.125711

[158] Cunningham BE, Siddiqui S, and Brander SM. How to Identify a Model Species. In Aquatic Ecotoxicology: Understanding Pollutants, Aquatic Organisms, and Their Environments. Cham: Springer International Publishing, 2024. p. 59–84. DOI: 10.1007/978-3-031-53130-9_5

[159] Chen CY, Lin Z. Exploring the potential and challenges of developing physiologically-based toxicokinetic models to support human health risk assessment of microplastic and nanoplastic particles. Environment International. 2024:108617. DOI: 10.1016/j.envint.2024.108617

[160] Kutralam-Muniasamy G, Shruti VC, Pérez-Guevara F, Flores JA. The emerging field of inductively coupled plasma mass spectrometry for (micro) nanoplastic analysis: "The 3As" analysis, advances, and applications. TrAC Trends in Analytical Chemistry. 2024:117673. DOI: 10.1016/j.trac.2024.117673

[161] Christopher EA, Christopher-de Vries Y, Devadoss A, Mandemaker LD, van Boxel J, Copsey HM, Dusza HM, Legler J, Meirer F, Muncke J, Nawrot TS. Impacts of micro-and nanoplastics on early-life health:

A roadmap towards risk assessment. Microplastics and Nanoplastics. 2024; **4**(1):13

[162] Kibria G. Exposure routes of microplastics (MPs) to humans and possible risks of MPs to human health from food and the environment: A short review. Journal of Food Safety and Hygiene. 2024. DOI: 10.18502/jfsh.v10i1.16440

[163] Morsella A, Castagna C, Messina R, Scarsi N, Ricciardi W, Cadeddu C. A systematic review on the relationship between marine pollution and human health in small island developing states. CABI One Health. 2025;**4**(1):0007. DOI: 10.1079/cabionehealth.2025.00

[164] Giroux MS, Siddiqui S, and Brander SM. Classic contaminants in aquatic ecosystems: POPs, PFAS, heavy metals, and microplastics. In: Siddiqui S, and Brander SM, eds. Aquatic Ecotoxicology: Understanding Pollutants, Aquatic Organisms, and Their Environments Cham:Springer International Publishing; 2024. p. 43–58. DOI: 10.1007/978-3-031-53130-9_4

[165] Shukla S, Pei Y, Li WG, Pei DS. Toxicological research on nano and microplastics in environmental pollution: Current advances and future directions. Aquatic Toxicology. 2024:106894. DOI: 10.1016/j.aquatox.2024.106894

[166] Teiba II, El-Bilawy EH, Abouelsaad IA, Shehata AI, Alhoshy M, Habib YJ, Abu-Elala NM, El-Khateeb N, Belal EB, Hussain WA. The role of marine bacteria in modulating the environmental impact of heavy metals, microplastics, and pesticides: A comprehensive review. Environmental Science and Pollution Research. 2024;1–34. DOI: 10.1007/s11356-024-35520-y

[167] Covello C, Di Vincenzo F, Cammarota G, Pizzoferrato M. Micro (nano) plastics and their potential impact on human gut health: A narrative review. Current Issues in Molecular Biology. 2024;**46**(3):2658–2677. DOI: 10.3390/cimb46030168

[168] Faburé J, Hedde M, Le Perchec S, Pesce S, Sucré E, Fritsch C. Role of trophic interactions in transfer and cascading impacts of plant protection products on biodiversity: A literature review. Environmental Science and Pollution Research. 2025;**32**(6):2993–3031. DOI: 10.1007/s11356-024-35190-w

[169] Rademaker M, van Leeuwen A, and Smallegange IM. Why we cannot always expect life history strategies to directly inform on sensitivity to environmental change. Journal of Animal Ecology. 2024;**93**(3):348–366. DOI: 10.1111/1365-2656.14050

Chapter 2

Nanoplastics in the Biological System: A Microscopic Menace with an Impact

John Dennis and Tapan Kumar Mistri

Abstract

Nanoplastics, ultra-small polymer particles measuring less than one micrometre, have become a widespread and mostly invisible environmental pollutant with significant biological consequences due to their high surface activity, mobility, and capacity to traverse biological barriers. Nanoplastics, on the other hand, can easily move through air, water, soil, and living things, which means that everyone is exposed to them. This is not the case with macro- and microplastics, which are usually only found in certain places. This chapter examines their origins, physicochemical characteristics, environmental dispersion, and biological impacts, emphasising their interactions within living organisms. Experimental results from cellular and animal studies indicate that nanoplastics induce oxidative imbalance, mitochondrial dysfunction, lysosomal disruption, inflammatory signalling, and cellular apoptosis across various biological systems, including neural, immune, reproductive, and epithelial tissues. Animal studies show more evidence of placental transfer, foetal accumulation, metabolic dysfunction, and long-lasting effects on brain development, which raises concerns about effects that could last for generations. Direct evidence of human exposure has been established by the identification of nanoplastics in organs and biological fluids, including blood, breast milk, brain tissue, and the placenta, suggesting systemic distribution via ingestion, inhalation, and dermal pathways. Nanoplastics are even more toxic because they actively bind to and carry dangerous substances like heavy metals and persistent organic pollutants, which makes cellular damage worse by delivering contaminants together. Even though there is more and more evidence that plastic is bad for the environment, rules still mostly only apply to bigger plastic items. This chapter discusses the limits of analysis, looks at recent advances in detection, and points out major gaps in our knowledge about long-term exposure and environmental change. It stresses the need for more focused research and government action.

Keywords: nanoplastics, bioaccumulation, inflammation, cell death, carcinogenicity

1. Introduction

The introduction of synthetic polymers in the early 20th century changed materials science. Plastics became essential in almost every area of modern life. Their light weight, strength, resistance to chemicals, and low cost have made them widely used in packaging, construction, electronics, agriculture, and healthcare [1]. As global plastic production surpasses 400 million metric tonnes each year, worries have grown about the environmental impact and biological risks of this common material [2]. Plastics, made to resist breaking down, build up in land and water ecosystems. Over time, they break into smaller pieces due to environmental factors like UV radiation, heat, and physical wear [3].

These fragments, once reduced to sizes smaller than 5 millimeters, are classified as microplastics. However, even smaller particles, called nanoplastics, usually defined as those smaller than 1000 nanometers, present the greatest risk to living systems [4]. Nanoplastics may originate directly from industrial applications, such as in drug delivery, cosmetics, and toiletries, or indirectly through the gradual breakdown of larger plastic debris in the environment [5]. Their small size and large surface area give them special properties that set them apart from larger macro- and microparticles. These properties include increased surface reactivity, stability in liquid, and the ability to cross biological membranes [6, 7, 8]. Once they enter the environment, nanoplastics show complex behavior. Their interactions with natural organic matter, inorganic ions, and other small particles can change their charge, water-repelling ability, and how they clump together [9]. These changes affect how they move in water, how they settle, and how available they are for biological use. Unlike microplastics, which can settle in sediments or get caught in living organisms, nanoplastics can stay suspended in the water or seep into tiny soil spaces. This could impact groundwater and farming systems [10]. The atmospheric transport of nanoplastics has also been confirmed, suggesting a global dispersion route even into remote regions such as Arctic ice and mountain air [11].

The spotting and estimation of nanoplastics present notable analytical challenges due to their small size, low mass intensity, and physicochemical similarities to natural particulates [12]. Conventional imaging techniques, including scanning electron microscopy (SEM) and transmission electron microscopy (TEM), do not offer chemical specificity. On the other hand, spectroscopic methods like Fourier-transform infrared (FTIR) and Raman spectroscopy require particle sizes that often exceed the dimensions of nanoplastics [13]. Progressive methods such as pyrolysis gas chromatography-mass spectrometry (Py–GC–MS), field-flow fractionation combined with multi-angle light scattering (FFF-MALS), and surface-enhanced Raman scattering (SERS) have begun to address this gap. However, standardized protocols for ecological and biotic samples are still under development [14]. The biological relevance of nanoplastics is underscored by their ability to traverse physiological barriers. Following environmental exposure, nanoplastics can enter organisms through oral, dermal, and nasal contact. Their nanoscale size enables them to bypass biological filters and penetrate systemic circulation [15]. Experimental studies have shown that nanoplastics can localize in several organs, such as the liver, brain, and placenta, highlighting their capacity to infiltrate critical tissues [16]. In aquatic organisms, consuming nanoplastics has been linked to problems with feeding, oxidative stress, and changes in reproductive function. For instance, filter-feeding invertebrates and fish have demonstrated accumulation of nanoplastics in their gastrointestinal tracts. This accumulation affects metabolism, enzyme activity, and behavior [17].

More concerning are the effects seen at the subcellular level in experiments involving both cells and animals [18, 19]. Cellular uptake of nanoplastics has been observed in human epithelial and immune cells. This often occurs through endocytosis or passive diffusion [20]. Once inside the cell, nanoplastics can settle in lysosomes, mitochondria, and even the nucleus. They may cause problems with mitochondria, lead to the creation of reactive oxygen species (ROS), and trigger inflammatory signaling pathways [21–23]. These processes could lead to genotoxicity, cell cycle arrest, or apoptosis. In addition, nanoplastics can absorb and transport harmful environmental agents like heavy metals, PAHs, and POPs. They effectively act as carriers for increased toxicity through a "Trojan horse" mechanism [22, 24]. The effects of these interactions go beyond the individual organism to larger ecological and intergenerational levels. In model vertebrate systems, maternal exposure to nanoplastics has caused them to move across the placental barrier and build up in fetal tissues. This suggests that developmental stages may be particularly vulnerable [25, 26]. Reduced fetal and placental weights, increased fetal resorptions, and changes in gene expression have been reported. These findings suggest a link between nanoplastics and reproductive toxicity, as well as developmental issues [26]. These effects raise concerns about heritable epigenetic changes. These changes could carry health problems through generations.

Ecosystem-level disruptions are becoming more noticeable. In aquatic food webs, nanoplastics that enter at the base through plankton or microbial ingestion can build up, accumulating in higher trophic levels [27]. This has the potential to disturb predator-prey dynamics, energy transfer, and population stability. In soils, nanoplastics may interfere with microbial diversity and function, inhibiting critical ecosystem services, notably nutrient cycling, organic matter degradation, and symbiotic plant-microbe interactions [28]. Their interactions with root systems and soil microbes suggest potential effects on farming output and plant health [29]. From a human health standpoint, exposure to nanoplastics is almost impossible to avoid because they are everywhere—in food, water, and air. Nanoplastics have been found in bottled water, seafood, table salt, and even in human blood and placenta [30–33]. Although direct evidence of human toxicity is limited, laboratory studies suggest that prolonged or repeated exposure could lead to chronic inflammation, endocrine disruption, neurotoxicity, and immune modulation. The cumulative burden of nanoplastics, especially in sensitive groups, including infants, pregnant women, or individuals with weakened immune systems, necessitates urgent investigation [34–37].

Despite the increasing evidence of their biological and ecological effects, nanoplastics are largely overlooked in current regulations. These regulations mainly target microplastics or large plastic waste. This oversight stems from the limited data on their environmental presence, toxicity levels, and ways people are exposed. Tackling this issue will require global cooperation, improved analytical tools, and thorough risk assessments. Nanoplastics do not just passively break down; they actively disrupt cellular and ecological systems. In this chapter, we will investigate their harmful effects on biological systems.

2. Physicochemical properties of nanoplastics

Nanoplastics have various chemical compositions. These compositions are influenced by how they are made and the changes they undergo in the environment. They originate either as primary nanoplastics, released directly in nanoparticle form, or as secondary nanoplastics, which are caused by the gradual fragmentation of bigger plastics and microplastics through mechanical wear, heat, chemical reactions, and biological activity [38–40]. Typical polymer types found consist of PS, PE, PP, PVC, PET, and PA [39, 41, 42]. Studies have shown that nanoplastics from consumer products like cosmetics, synthetic textiles, and food packaging often consist of PE, PS, PET, or silicone-based polymers. Each of these has its own unique structural features [39, 43, 44].

The structural characteristics of nanoplastics are significantly influenced by both standardized production methods and environmental degradation. In the laboratory, controlled synthesis yields highly uniform, contaminant-free nanoparticles with molecular weight distributions and crystallinity comparable to commercial-grade plastics, providing essential materials for environmental monitoring and toxicological research [42]. In natural environments, however, processes like photo-oxidation, abrasion, and biofouling cause chemical changes, adding reactive species such as OH or COOH, shortening polymer chains, and even leading to mineral deposition, as seen in calcium carbonate ($CaCO_3$) accumulation on aged PET nanoparticles [45–47]. These changes make environmentally weathered nanoplastics more hydrophilic, chemically reactive, and often less negatively charged. Field observations in both living organisms, such as mussels, and non-living samples show that nanoplastics exist in a range of polymer types. Each type has its own unique shape and chemical profile, which can be identified using Raman imaging or mass spectrometry at nanometer scales [43, 48, 49]. Fragments as small as 24 to 52 nm have been found in commercial scrubs. This shows their ability to break down and their common presence around the world [44]. It is estimated that in certain areas, seafood consumption may add more than 2 mg per person each year [48]. Such findings emphasize the need to classify nanoplastics based on their polymer type, how they age in the environment, and changes to their surface, such as oxidation or eco-corona formation. These factors affect how long they last, how they interact, and their toxicity [45–47].

Their high surface-area-to-volume ratio, combined with surface heterogeneity and reactive sites formed during weathering, gives nanoplastics a strong capacity to adsorb pollutants [46, 47, 50]. These nanoplastics act as mobile carriers, binding and transporting contaminants that influence environmental fate and biological exposure. Lipophilic pollutants such as PAHs, PCBs, and endocrine disruptors adhere to their surfaces via hydrophobic interactions. Weathering slightly reduces hydrophobic binding but introduces functional groups that enhance adsorption [46, 47]. An eco-corona of organic matter (lamina) forms spontaneously on nanoplastics, altering sorption by adding or masking binding sites. Heavy metals bind mainly through ionizable groups like carboxyl (-COOH) or amine ($-NH_2$), introduced by aging or modification [45, 47, 51]. These groups enable electrostatic attraction, complexation, or ion exchange. Experimental evidence shows that polystyrene nanoplastics can lead to higher accumulation and toxicity of gold ions in zebrafish embryos [52]. In environmental and biological contexts, nanoplastics also adsorb to form an active protein corona. The composition of this corona depends on particle chemistry and aging. This corona influences how cells recognize and take up the particles. Aged particles are often taken up more easily by macrophages [45, 53]. Adsorption involves forces such as Van der Waals, hydrophobic partitioning, hydrogen bonding, electrostatic and cation–π interactions, and sometimes covalent bonding from environmental processes. Surface charge and functionalization govern

contaminant binding and cellular interactions, with associations detected at concentrations as low as 10^{-7} g/L [47, 53, 54], confirming their role as highly effective contaminant vectors.

Nanoplastics are continuously generated in the environment from the fragmentation of larger plastics through erosion, ultraviolet deterioration, and biological behavior [39, 47, 50]. Controlled experiments as well as field surveys show that ordinary actions, such as washing 1 g of polyester fleece, can release over 3.3×101110^{11} particles (2.1 mg) per wash [39]. Once released, their dispersion is regulated by particle size, compactness, surface composition, and interactions with organic matter and minerals, which can promote aggregation or sedimentation [47, 50, 55]. Techniques like filtration, Raman spectroscopy, and single-particle inductively coupled plasma mass spectrometry have confirmed their presence in various environmental samples. The highest concentrations usually fall within the 20 to 200 nm range [48–51]. Environmental modeling shows that these particles are common and long-lasting. They also have a high potential for bioaccumulation and trophic transfer.

Within biological systems, nanoplastics are easily absorbed through barriers like plant roots, skin, mucous membranes, and even cellular membranes [38, 56, 57]. Laboratory experiments have confirmed that aquatic species can ingest and accumulate substances in their tissues. This translocation occurs in edible tissues, allowing humans to come into contact with these substances through the food chain [48, 52]. Inside cells, uptake happens mainly through endocytosis or phagocytosis. This process is influenced by particle size, surface chemistry, and corona characteristics [45, 56, 58]. It can damage tissue integrity, change gene expression, affect immune responses, and cause inflammation or oxidative stress [59]. Prolonged low-level exposure has been connected to lasting biological effects, including the activation of stress-response genes, indicating ongoing risks to ecosystems [54, 60]. Simultaneous exposure to other pollutants can increase toxic effects. Nanoplastics enhance the absorption and availability of both metallic toxins and organic compounds [46, 52].

Measuring nanoplastics in complex samples is still difficult because of their tiny size and variation. Recent advances in surface-enhanced Raman spectroscopy, single-particle ICP-MS with a limit of sensing as low as 8.4×10^5 particles/L, and high-resolution electron microscopy have made it possible to detect and characterize them at very low concentrations [41, 43, 49, 51]. These analytical improvements are critical for quantifying their environmental distribution, accumulation, and possible dangers to humans and ecosystem conditions.

3. Ubiquity of micro and nanoplastics in the human body

Studies indicate an extensive buildup of nanoplastics in human tissues and fluids, ranging from submicron to microplastic dimensions. They manifest as pieces, fibers, beads, granules, or shards, encompassing polymers such as polyethylene (PE), polypropylene (PP), polystyrene (PS), polyvinyl chloride (PVC), polyethylene terephthalate (PET), and synthetic composites (**Table 1**). A recent review has confirmed the presence of these particles in many human systems, including the cardiovascular, digestive, endocrine, integumentary, lymphatic, respiratory, reproductive, and urinary systems. They have also been found in biological materials such as breast milk, meconium, semen, stool, sputum, and urine [61]. The

Sample/ Organ	Number of Samples	Particle Size Range	Particle Shape	Main Polymer Types	Reference
Blood	22	0.7–1.5 μm	Not specified	PET, PE, PS, & PMMA	[62]
Kidney (post-mortem)	~50	<1 μm (NPs predominate)	Shard-like	PE, others	[64]
Liver (post-mortem)	~50	<1 μm to several μm	Shard-like	PE, PP, & PVC	[64]
Heart (right atrium)	15	20–500 μm	Irregular	PET, PS, & PE	[62]
Carotid plaque	304	NPs and MPs (sub-micron–μm)	Jagged/ irregular	PE, & PVC	[65]
Saphenous vein	5	≥5 μm	Irregular	Alkyd resin, polyester blends	[62]
Testes	23	0.28– > 1 μm	Irregular	PS, PE, others	[66]
Placenta	6	≈5–10 μm or smaller	Irregular	PP, PE, & PVC	[66]
Breastmilk	34	2–12 μm	Fragment, fiber, bead, granule	PE, PVC, & PP	[74]
Sputum	22	20–500 μm (median: 75.43 μm; IQR: 44.67–210.64 μm)	Fiber, fragment, irregular	PU (dominant), PES, CPE, and AV	[67]
Urine	6	4–15 μm	Irregular fragments	PEVA, PVC, PP, & PE	[75]
Olfactory bulb (human brain)	15	Not explicitly stated (MPs, not NPs).	Particles, fibers	PP	[63]
Semen	10	16 Particles	2-6 μm	PP, PE, PET, PS, PVC, PC, and POM	[76]
	25	0.23 ± 0.45 MPs/mL	21.76–286.71 μm	PVC, PE, PA, PP, PS, and PET	[77]
Human vitreous humor	49	8,543 particles in total	<50 μm	PA-66, PVC, & PS	[78]
Bone marrow	16	51.29 μg/g	<100 μm	PE, PS, PVC, PP, and PA-66	[79]
Human endometrium	20	21 MPs/100 mg	20-500 μm	ACR, PE, PET, PP, PS, PU, PVC, BR, CPE, EAA, COP, FKM, EVA, & CO.	[80]
Uterine fibroids & Myometrium	48	1.5 ± 1.17 MPs/g	2.79–22.79 μm	PE, PP, PEVA, PE–co–PP, and ABS	[47]
Penile tissue	6	46 in total	20–500 μm	PET, PP, PMMA, PS, PVC, PTFE, and PU	[81]

Table 1.
Detection of Microplastics (MPs) and Nanoplastics (NPs) in human biological tissues, organs, and fluids. *The table summarizes studies that report on sample type, the number of samples analyzed, particle size range, morphology, and the main types of polymers found.*

identification of microplastics in human saphenous vein tissue using μFTIR spectrometry [62] provides clear evidence of their ability to reach vascular structures, suggesting systemic dissemination via the circulatory network.

One of the most striking recent findings is the detection of microplastics in the human brain. Polypropylene particles were found in the olfactory lobe of eight out of fifteen deceased individuals. This suggests that the olfactory pathway acts as a direct route for exposure to the central nervous system [63]. Complementary studies using pyrolysis-GC/MS and electron microscopy have confirmed the presence of micro and nanoplastics in the brain, kidney, and liver of cadavers. Polyethylene is the most common polymer found in brain tissue, and concentrations generally increase with age. In individuals with dementia, particles have been observed in the vascular walls of the brain and in immune cells. This suggests a possible connection to neuropathological issues [64].

Within the circulatory system, micro and nanoplastics have been found in venous blood. The main types detected are PE, PET, PS, and PMMA. These particles range in size from 0.7 μm to 1.5 μm [62]. Their presence in blood highlights the potential for systemic distribution and reaching peripheral tissues. Post-mortem studies show that both nanoscale and micron-sized PE particles build up in the kidney and liver, with levels rising as time goes on [64], indicating that micro and nanoplastics can cross biological barriers and remain in metabolic organs. This raises concerns about long-term toxicity specific to those organs.

Cardiovascular tissues are another major site for accumulation. In patients who have undergone heart surgery, microvesicles composed of PET, PS, and PE (20–500 μm) have been found in the saphenous vein, heart muscle, and carotid plaque. Their presence is associated with local inflammatory responses [62]. PE and PVC found in carotid plaques are associated with a significantly higher risk of adverse cardiovascular events, with a hazard ratio of 4.53 [65].

The reproductive system is also impacted. Testicular tissue contains PS and PE particles that are about 0.28 μm or larger, with higher amounts of these particles correlating with lower sperm counts [66]. Studies have confirmed the presence of microplastics in the placenta, covering both maternal and fetal regions. This includes PP, PE, and PVC. These findings show translocation across the placental barrier and emphasize the chance of direct exposure to the fetus.

The wide distribution of nanoplastics throughout the human body shows several ways they can enter, mainly through nasal and oral exposure. Dermal absorption occurs less often but is still a possible route. We can confirm respiratory uptake because nanoplastics have been found in human sputum. The median particle size is 75.43 μm, and 21 different polymer types were identified in all the samples analyzed [67]. Exposure introduces nanoplastics through contaminated food, water, and packaging. Dermal contact, while not as well understood, is supported by observations in occupational and cosmetic settings [68]. Once internalized, micro and nanoplastics can cross epithelial and endothelial barriers via paracellular transport, endocytosis, or persorption, enabling their dissemination through the blood and lymphatic systems [69, 70]. Their repeated detection in the liver, kidney, and brain demonstrates an ability to overcome physiological defenses and accumulate in tissues with slow clearance rates [31, 71, 72]. Although elimination is possible, as evidenced by the presence of microplastics in urine and stool, retention in critical organs suggests potential for long-term bioaccumulation. Maternal transfer has also been confirmed, with microplastics found in 26 of 34 breastmilk samples, predominantly PE, PVC, and PP [73], indicating that exposure is effectively unavoidable under current environmental conditions.

4. Reports based on in-vitro analysis

Polystyrene nanoplastics (PS–NPs) cause various effects on cells, which depend on their size, surface charge, concentration, and duration of exposure (**Figure 1**). Nanoplastics, whether pristine, weathered, surface-modified, or contaminated with toxins such as lead, disturb cellular homeostasis by inducing oxidative stress, mitochondrial dysfunction, altered autophagy, inflammation, DNA damage, metabolic alterations, and defective signaling across several cellular models (**Table 2**). Their uptake and effects have been confirmed in different human and animal cell lines. For example, Caco-2 intestinal epithelial cells take up PS–NPs in a concentration-dependent manner. Significant uptake starts at 25 μg/mL and can reach 86% at 100 μg/mL, which also leads to a threefold increase in fluorescence intensity [82]. Similarly, exposure to 50 nm PS–NPs for both short-term (24 hours) and long-term (8 weeks) periods led to significant internalization, including occasional nuclear localization. However, even with this uptake, the genotoxic response was minimal. There was only slight overexpression of oxidative stress-related genes (HO-1 and SOD2) and mild mitochondrial changes. This suggests low overall cytotoxicity under the tested conditions [83].

The harmful effects of PS-NPs depend heavily on their size. Small particles (<100 nm) are usually more dangerous than larger ones. This is evident in liver LO2 cells, which took in more PS-NPs than kidney 293 T cells but showed greater resistance to toxicity. When lead ions (Pb^{2+}) attach to PS-NPs, they increase the harmful effects by raising oxidative stress, causing mitochondrial dysfunction,

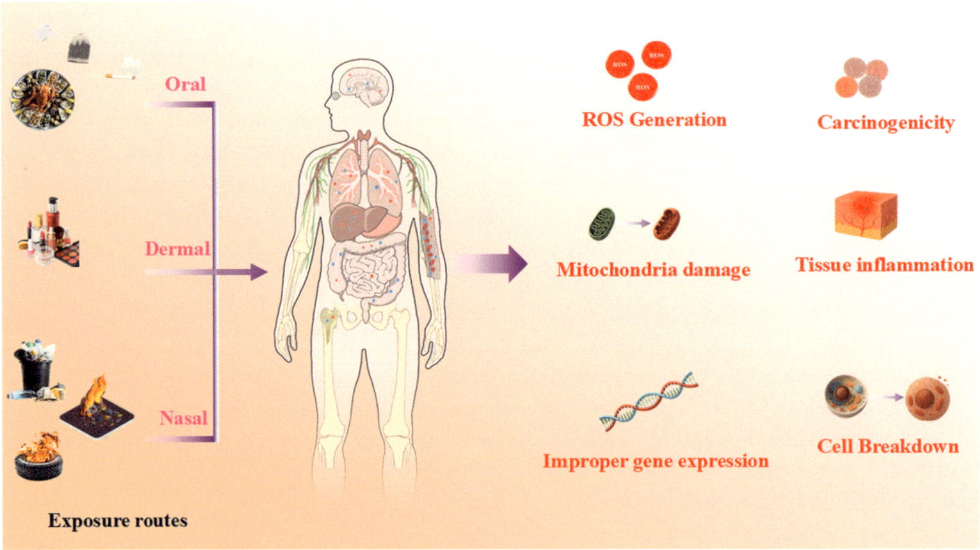

Figure 1.
Schematic illustration of potential exposure routes and associated biological effects of nanoplastics in the human body.

S. No	Model	Cell type	NPs	NPs size	Concentration of NPs	Time of exposure	Effects	Reference
1	Human colon cells	Caco-2	Fluorescent PSNPs & non fluorescent PSNPs	50 nm	0.0006, 0.26, 1.3, 6.5 mg/cm^2	24 h (Short term)	Only a few mitochondrial alterations were detected at higher concentrations.	[83]
					0.26 mg/cm^2	8 weeks (long term)	Slight overexpression of *HO1* and *SOD2* genes	
2	Human alveolar cells	A549	Amine modified	20 nm	10, 20 and 40 µg/mL	24 h	Induce epithelial mesenchymal transition, ATP reduction, & Mitochondrial dysfunction.	[85]
			unmodified PSNPs	20 nm	10, 20 and 40 µg/mL			
				50 nm	40, 80 and 160 µg/mL			
3	Human granu-losa cell line	KGN	PSNPs	15 nm	100 µg/mL	48 h	Decreased cell viability, Upregulation in genes involved in ROS generation & down regulation in ROS scavenging genes, *NOX4* & *GPX1* upregulation leads to oxidative stress in the presence of salidroside.	[86]
				38 nm				
4	Human embryonic cells	293 T	PSNPs + Pb^+	<100 nm	1, 5, 25, 75, and 125 µg/mL	24 h	Intracellular ROS production, Cell membrane damage, ATP reduction, Mitochondria damage, & Cell mortality	[84]
	Human cells	LO2	PSNPs + Pb +	20 nm	1, 5, 25, 75, and 125 µg/mL	24 h		
5	Human liver cells	HepG2	Aminated and non aminated PSNPs	100 – 5,000 nm	10 – 100 µg/mL	4 – 24 h	Aminated PSNPs reduced cell viability with increase in concentration, Cell apoptosis, & cause inflammation by upregulating IL8.	[89]
6	Human colon cells	Caco-2	PSNPs & Fluorescent PSNPs	0.04 to 1 µm	175 and 200 µg/mL	12 & 24 h	Less cytotoxicity	[82]

S. No	Model	Cell type	NPs	NPs size	Concentration of NPs	Time of exposure	Effects	Reference
7	Human umbilical vein endothelial cells	HUVECs	PSNPs	100 & 500 nm	0, 5, 10, 25, 50 and 100 μ g/mL	24 & 48 h	Cell viability decreased with a increase in concentrations, Conversion of gene LC3B I to LC3B II, Upregulation in *BECN1*, Lysosomal damage, & autophagy	[20]
8	Human liver cells	HepG2	Fluorescent PSNPs, PSNPs-COOH, & PSNPs-NH_2	50 nm	<100 μ g/mL	24 – 48 h	Increase in malondialdehyde, Decrease in *SOD*, inducing oxidative stress, DNA damage & lipid oxidation	[23]
9	Human nasal epithelial cells	HNEpC	PETNPs (From drinking bottle)	<100 nm	<100 μ g/mL	24 h	Increase in ROS, & Damage in mitochondria.	[21]
10	Human gastric cells	SNU-1	Aminated, carboxylated and non-functionalised PSNPs	50, 100, & 1000 nm	0.1–100 μ g/mL	24 h	Surface charged induced cell death, & Both apoptosis and necrosis happened	[22]
11	Human nasal & airway epithelial cells	HNEpCs & HEPES	PSNPs	50 & 100 nm	100 μg/mL	24 h	Loss of Mitochondrial Membrane Potential, Generation of iRyOS, increase in expression of LC3–II, & defective autophagy	[24]
12	Human intestinal cells	Caco-2	PET & PSNPs	<800 nm	10^2-10^6 particles/mL	48 h	Glycolytic capacity increased, Increase in basal glycolysis, Elevated cellular stress, & increase in Mito ATP	[88]
13	Adipose stromal cells	ASCs	PSNPs	<100 nm	5, 25 and 75 μg/μL	48 h	Generation of free radicals, PAI1 was inhibited, & TNFα was stimulated	[34]
14	Human neural stem cell	hNS1	PSNPs	30 nm	0.5, 2.5 and 10 μg/mL	4 days	DNA damage, alterations in inflammatory response, andapoptosis	[35]
15	Jejunal porcine epithelial cell line	IPEC-J2	PSNPs	100 nm	5.47 *10^{11} Particles/mL	5-60 h	Oxidative and genotoxic stress, alter signalling pathways, and activate stress related apoptotic responses, and threatening cellular homeostasis.	[87]

S. No	Model	Cell type	NPs	NPs size	Concentration of NPs	Time of exposure	Effects	Reference
16	Breast cancer	SK–BR-3	Unmodified PSNPs &	0.1 μm	12.5 to 200 μg/ mL	72 h	Sticking to the cell membrane, & Inhibit cell signalling pathway	[37]
	Fibroblasts	NIH3T3	Amine modified PSNPs					
17	Human cells	Placental primary cells	PSNPs	20 & 100 nm	0.1, 10 & 100 μg/ mL	48 h	Impaired endocrine activity, Alterations in mitochondrial function, lysosomal activity, and autophagy	[36]

Table 2.
Summary of in vitro studies reporting Nanoplastic (NP) accumulation and effects in various human and mammalian cell types. *The table compiles key experimental parameters, including NP type, particle size, concentration, exposure duration, and observed cellular responses.*

and depleting ATP [84]. In human placental trophoblasts, 20 nm particles resulted in more severe cytotoxicity compared to 100 nm particles, even at very low doses (1 μg/mL). They induced oxidative stress, caused the release of inflammatory cytokines (IL-6, TNF-α), and disrupted endocrine activity, such as reducing human chorionic gonadotropin secretion [36]. A similar pattern was observed in endothelial cells, where minute particles (100 nm) were absorbed and triggered autophagy initiation, whereas larger particles (500 nm) localized on the cell membrane, causing membrane damage and lactate dehydrogenase release [20].

Surface charge and functionalization further modulate the cytotoxicity of nanoplastics. Amino-functionalized PS-NPs displayed the most severe effects on human liver cells, causing dose-dependent reductions in cell viability and oxidative damage. Initially, low doses triggered an increase in antioxidant defenses such as superoxide dismutase (SOD) and glutathione (GSH), but prolonged or high-dose exposure depleted these defenses, resulting in increased lipid peroxidation indicated by malondialdehyde accumulation [23]. Similarly, positively charged amino-functionalized 20 nm PS-NPs strongly reduced cell viability and promoted induction of epithelial-mesenchymal transition (EMT) through the downregulation of E-cadherin and the upregulation of *MMP2*. These effects were linked to oxidative stress driven by *NOX4* overexpression and were accompanied by mitochondrial dysfunction characterized by decreased ATP production and altered membrane potential [85]. In gastric cells, smaller aminated particles demonstrated higher uptake and induced rapid apoptosis-necrosis within four hours, whereas carboxylated and unmodified particles caused delayed cytotoxicity [22]. Interestingly, protein corona formation around PS–NH_2 particles significantly reduced toxicity and restored normal signaling through the MAPK/ERK and PI3K/AKT pathways, demonstrating a protective role of biomolecule adsorption [37].

Disruption of key signaling pathways has been documented in several cell types. In human granulosa cells, PS-NPs at 100 μg/mL inhibited proliferation and triggered apoptosis by inducing oxidative stress and dysregulating the Hippo signaling pathway through activation of *MST1, LATS1*, and *YAP1*, while downregulating *CTGF* and *CYR61*, which are critical for cell survival [86]. In IPEC-J2 cells, PS-NPs disrupted the *NOTCH* signaling pathway. They downregulated *NOTCH3* and activated *TP53*, leading to changes in the cell cycle checkpoints and causing genotoxic stress, as shown by the upregulation of *GADD45A* [87]. In neural stem cells, nanoplastics caused widespread changes in gene activity, including an increase in oxidative stress markers like *SOD1* and *CAT*, and in DNA repair genes such as *GADD45A* and *RAD51*. In contrast, the mismatch repair gene XRCC1 showed a decrease in activity. Apoptotic regulators, including *BCL2* and *HSP27*, rose in levels, accompanied by early activation of *CASPASE-7*. Inflammatory markers showed complex behavior that depended on the dose [35]. *HSP27* upregulation has been linked to neurodegenerative pathways, connecting nanoplastic exposure to possible neurological risks.

Mitochondrial dysfunction and problems with energy metabolism appeared as common effects in multiple studies. Exposure to nanoplastics led to a loss of mitochondrial membrane potential in nasal epithelial cells. This was linked to an increase in intracellular reactive oxygen species (iROS) and signs of faulty autophagy (LC3–II and p62). These findings suggest a problem with autophagic flux [21]. Similar findings were observed in airway epithelial cells. The increase in LC3–II indicated a problem with autophagy, even though the overall cell damage was quite low [24]. Additionally, weathered nanoplastics altered the metabolism in Caco-2 cells from mitochondrial respiration to glycolysis. This shift likely occurred because of electrostatic interference with mitochondrial function. PET nanoplastics also increased the basic glycolytic activity, which suggests a disruption in energy balance [88].

Inflammatory and oxidative responses were consistently observed in several models. HepG2 hepatocytes exposed to aminated PS-NPs exhibited increased *CASPASE-3* activity and *IL-8* expression after prolonged exposure. This suggests that oxidative stress can lead to cell death and inflammation [89]. In swine adipose stromal cells, prolonged exposure to PS-NPs resulted in oxidative imbalance. This was characterized by increased reactive oxygen species (O_2^-, H_2O_2, NO), reduced non-enzymatic antioxidant capacity, and alterations in inflammatory markers, such as elevated TNF-α and decreased PAI-1 [34]. PS-NPs disrupted critical endocrine and metabolic pathways in placental cells, underscoring their potential to harm essential physiological processes [36].

Collectively, these findings show that nanoplastic toxicity depends heavily on the context. It is influenced by physical and chemical properties, such as size and surface chemistry, along with the duration and amount of exposure. Some cell types are resilient, while others show significant oxidative stress, mitochondrial dysfunction, changes in signaling pathways, and inflammation. These effects can lead to cell death, poor barrier function, and changes in metabolism. These lab observations emphasize the urgent need for systematic studies that include realistic environmental concentrations, long-term exposures, and complex mixtures to more accurately predict risks to human health.

5. Reports based on in-vivo analysis

Many studies confirm nanoplastic (NP) translocation across the placental barrier, accumulation in multiple fetal organs, and intergenerational effects. Together, these findings highlight the capacity of nanoplastics to cross biological barriers, disrupt developmental processes, and induce multi-organ toxicity in a size-, dose-, and sex-dependent manner (**Table 3**). Emerging evidence highlights the severe implications of maternal exposure to polystyrene nanoplastics (PS-NPs) and microplastics (PS-MPs) during gestation and lactation on fetal development, organ function, and long-term health. These effects manifest through placental transfer, metabolic disruption, and neurodevelopmental alterations, with mechanisms often linked to oxidative stress, inflammation, and ferroptosis.

Transplacental passage of PS-NPs is well documented. Fluorescent imaging confirms their presence in maternal organs and various fetal tissues, including the brain, within 24 hours of exposure [90]. This ability to cross biological barriers highlights the risk to the developing fetus, especially because the blood-brain barrier is still immature. When mothers are exposed to nanoplastics, it consistently results in restricted fetal growth, reduced placental weight, and higher resorption rates [90–92]. Placental dysfunction is key to these outcomes. PS-NPs change important pathways that control blood clotting, complement activation, and cholesterol metabolism. This disrupts nutrient transport and energy balance. Metabolic profiling of placental tissue showed lower levels of sucrose and glycogen.

S. No	Animal model	Nanoplastics	Size	Concentration of exposure	Route	Time	Abundance	Effects	Reference
1	Sprague Dawley rats	Rhodamine-labeled PSNPs	20 nm	2.64 × 1,014 particles	Direct ingestion	GD 19-GD 20	2.27 × 1,017	Reduced Fetal and Placental Growth, Maternal NP Distribution, Fetal NP Accumulation, and Placental Transfer Confirmed by Perfusion.	[90]
2	Kunming mice	PSNPs	100 nm	0.1,1,10 mg/L	Liquid feed	7 week old	1.82 × 1,015	Growth impairment, Hepatic toxicity, reproductive toxicity, and oxidative damage in testis.	[67a]
3	C57BL/6 J mice	PSNPs	80 nm	1,5,25 µg/µL	Oral ingestion		3.55×10^{15}	Reduced birth weight in female offspring compared to males, intestinal structural changes, oxidative stress markers, ferroptosis indicators, and sex-specific toxicity.	[105]
4	CD-1 mice	PSNPs	5 µm,50 nm	10^6 ng/L	Oral ingestion	GD 0.5 – GD 17.5	1.46×10^{10}, 1.46×10^{16}	Fetal Growth and Placental Dysfunction, Hemodynamic Alterations, Middle Cerebral Artery Pulsatility Index, Fetal Hypoxia Indicators & Size-Dependent Severity.	[93]
5	C57BL/6 mice	PSNPs	100 nm	1, 10 mg/L	Oral ingestion	GD 0 – GD 17	1.82×10^{15}	Reduced fetal weight, disrupted cholesterol metabolism, metabolic disorders, and gene regulation affecting muscle development.	[25]
6	SD rats	PSNPs	25 nm,50 nm	0.5,2.5,10,50 mg/kg	Oral gavage	GD 0.5–GD 17.5, GD 0.5–GD 21.5	1.16×10^{17}, 1.46×10^{16}	Crossed the placental barrier; accumulation in the cerebellum, hippocampus, striatum, and prefrontal cortex; myelin impairment; structural and cellular effects; and motor deficits in offspring.	[99]
7	CD-1 mice	PSNPs	50 nm	10^6 ng/L	Liquid feed	GD 0.5 to GD 17.5	1.46×10^{16}	Altered brain metabolism, sex-specific effects, and abnormal fetal brain metabolism.	[26]
8	SD rats	PSNPs	50 nm	2.5 mg/kg/day	Oral gavage	GD 0.5 – PND 22	1.46×10^{16}	Accumulation in breast milk, reduced expression of neural developmental proteins, ferroptosis pathway, and cognitive deficits.	[100]

S. No	Animal model	Nanoplastics	Size	Concentration of exposure	Route	Time	Abundance	Effects	Reference
9	SD rats	PSNPs	50 nm	2.5 mg/kg/day	Oral ingestion	Gestational, lactational exposure	1.46×10^{16}	Altered monoamine neurotransmitters, reduced cortical thickness, disordered neocortical migration, and reduced postsynaptic density.	[101]
10	C57BL/ 6Jmice	PSNPs	40, & 193 nm	5,10 mg/L	Liquid feed	Dams: GD 9.5 – PND 28.5, Offspring: PND 28.5 – PND 113	2.84×10^{16}, 2.53×10^{14}	Disruption of Brain Development, Behavioral Alterations, and Potential Neurotoxicity.	[102]
11	Kunming mice	PSNPs	100 nm	0.1,1 and 10 mg/L	Liquid feed	GD0toPND21	1.82×10^{15}	Growth Impairment, Hepatic Toxicity, Reproductive Toxicity, and Testicular Oxidative Damage.	[67b]
12	C57BL/ 6Jmice	PSNPs	71.09 ± 6.63 nm	1,5,25 μg/μL	Oropharyngeal aspiration	Gestational exposure	5.30×10^{15}	Hepatic steatosis Both in DAMS and female offspring.	[94]
13	ICR mice	PSNPs	0.5, 5 μm	1,001,000 μg/L	Liquid feed	GD 0 – PND 0	1.46×10^{13} 1.46×10^{10}	Lipid and energy metabolism disruption, amino acid and carnitine metabolism alterations, gene expression changes, and maternal metabolism.	[95]
14	C57BL/ 6Jmice	PENPs	100 nm	2 mg/ L	Oral ingestion	Paternal exposure	2.52×10^{14}	Growth and Hormonal Alterations, Reproductive Toxicity, Epigenetic Changes, Gut Microbiota Alterations, and Intergenerational Effects.	[98]
15	C57BL/ 6Jmice	PSNPs & PPNPs	50 nm & 500 nm	0.5,10,100, 500 μg/day	Oral feed	GD 8 – PND 14	1.68×10^{13}	Increased offspring body weight, lipid metabolism dysregulation, gut microbiota alterations, direct NP accumulation in offspring intestine, and disruption of lipid metabolism and microbiota.	[96]
16	C57/ BL6Jmice	PSNPs	80 nm	1,5,25 μg/μL	Oropharyngeal aspiration	GD 0 – GD 21	3.55×10^{15}	Altered Birth Weight, Increased Adult Body Weight, Ferroptosis Activation, and Sex-Specific Toxicity.	[105]

S. No	Animal model	Nanoplastics	Size	Concentration of exposure	Route	Time	Abundance	Effects	Reference
17	ICR mice	PSNPs	0.5 μm	0.5,5,50 mg/L	Drink	GD 1 – PND 35 & PND 70	1.46×10^{13}	Testicular Development Disorders, Spermatogenesis Dysfunction, Hormonal and Cell Cycle Disruption, Meiotic Process Disturbance, Immune and Inflammatory Response, and Male Subfertility	[77]
18	ICR mice	PSNPs	1 μm	1.357 ng/g/day	Gavage	F0 PND 0 – PND 21	1.82 × 1,012	Delayed Puberty, Disturbed Estrous Cyclicity, Reduced Fertility, Elevated Testosterone Levels, Abnormal Follicle Development. Disrupted Ovarian Steroidogenesis & Ovarian Inflammation	[97]
19	ICR mice	PSNPs	1 μm	1.36×10^{-6}, 1.36×10^{-3} mg/g/day	Intragastric administration	GD 0 – PND 21	1.82×10^{12}	Increased Spleen Weight, Higher Number of Immune Cells, Increased Immune Cell Counts, and Altered Immune Balance.	[104]
20	ICR mice	PENPs	16.9 ± 1.9 μm	0.125,0.5,2 mg/day	Oral ingestion	90 days of parental exposure and to dams until lactation.	4.17×10^{8}	Reduced Body Weight Gain, Immune System Alterations, and Tissue Changes.	[103]
21	C57BL/6mice	PSNPs	100 nm	1,10 mg/L	Liquid feed	GD 0 to GD 17	1.82×10^{15}	Reduced Fetal Weight, Cellular Abnormality, Disturbance in Cholesterol Metabolism, and Altered Gene Expression	[92]
22	SD rats	PSNPs	1 μm	2 mg/kg/day	Oral ingestion	28 days	1.82×10^{12}	Shortened tibial length, altered calcium and phosphorus metabolism, upregulation of ER stress-related genes, and chondrocyte death increased.	[99]
23	SD rats	PSNPs	20 nm	2.64×10^{14} particles	Intratracheal instillation	GD 19 -GD 20	2.27×10^{17}	Vascular Dysfunction, Reduced Fetal & Placental Weight, & Cardiovascular Dysregulation in Fetuses	[106]

Table 3.
Summary of in vivo animal studies investigating nanoplastic (NP) exposure, biodistribution, and biological effects. *The table compiles data from rodent models exposed to polystyrene nanoplastics (PS-NPs) and polyethylene nanoplastics (PE-NPs) of varying sizes, concentrations, and functionalization via different exposure routes (oral ingestion, gavage, liquid feed, oropharyngeal aspiration, and intratracheal instillation).*

In fetal tissues, genes related to muscle development and fat processing were downregulated. This was also linked to disrupted PPAR signaling and inflammatory activation through the JAK-STAT pathway [92]. Physiological effects go beyond just growth restriction. Size-related changes in fetal blood flow have been observed. Nanoplastics reduced blood flow in the umbilical artery and the cerebral pulsatility index. This triggered a brain-sparing response, which is a sign of fetal hypoxia [93]. Cardiovascular changes in offspring include higher blood flow velocity in major arteries and reduced vascular compliance. These changes relate to issues in maternal blood vessels, such as increased heart weight, problems with the endothelium, and increased smooth muscle reactivity after exposure to PS-NPs in the lungs [91]. These findings suggest a dual maternal-fetal cardiovascular risk that may predispose to long-term disease.

Organ-specific toxicities are also clear. Maternal PS-NP exposure during pregnancy and breastfeeding causes liver damage in offspring. This damage is marked by reduced liver weight, oxidative stress, inflammatory infiltration, and disrupted glycometabolism [67]. High-dose inhalation exposure in mothers worsens these effects. It increases lipid accumulation and leads to NAFLD-like symptoms by boosting lipid synthesis pathways. Female offspring are particularly vulnerable [94]. Microplastics disrupt lipid balance in a way that depends on sex. They change triglyceride and cholesterol levels and reduce the activity of genes related to β-oxidation in offspring [95]. These metabolic issues show that there could be a risk of obesity and metabolic disorders being passed down through generations. This is backed by evidence of changes in breast milk fat composition and shifts in gut bacteria that promote obesity [96].

Reproductive toxicity is another major concern. Male offspring exposed to PS-NPs show reduced testis weight, disruption of the seminiferous epithelium, and lower sperm counts. These effects are caused by oxidative injury and hormonal imbalance [67, 77]. Microplastics also disrupt sperm production and trigger immune-related responses. Additionally, exposure during breastfeeding causes ovarian problems and produces reproductive effects that can impact future generations [97]. Paternal exposure to polyethylene nanoplastics adds another layer of complexity. It leads to intergenerational effects tied to changes in gene expression and imbalances in gut microbiota. Both factors contribute to reduced fertility [98].

Neurological effects from prenatal PS-NP exposure are significant and varied. Nanoplastics build up in important brain areas, such as the cerebellum, hippocampus, and prefrontal cortex. This results in hypomyelination, death of oligodendrocytes, and problems with motor coordination [99]. Metabolomic changes involving lower levels of GABA, creatine, and glucose suggest problems with neurotransmission and energy regulation. There are also differences in amino acid metabolism based on sex [26]. Proteomic analyses show a decrease in synaptic and cytoskeletal proteins, including MAPT and KIF21A. This is linked to the activation of ferroptosis and neuroinflammation [100]. Structural changes in the brain, like cortical thinning, abnormal neuronal migration, and synaptic deficits, are linked to neurochemical imbalances. These include changes in dopamine, GABA, and glutamate levels. As a result, individuals may experience behavioral issues such as hyperactivity and memory problems [101]. Consistently, exposure to nanoplastics disrupts Wnt and Shh signaling, Gabra2 expression, and neuronal shape. This reinforces their role in neurodevelopmental disorders [102].

Finally, the systemic nature of nanoplastic toxicity affects bone growth and immune function. Microplastics hinder endochondral ossification through ER stress pathways, which reduces tibial length and disrupts growth plate organization [99]. Immune dysregulation includes more B cell and T cell subsets in offspring. This generational persistence suggests epigenetic reprogramming [103, 104]. Collectively, these findings show that exposure to nanoplastics in mothers poses a significant environmental risk, with effects that can span generations. The mechanisms involved include oxidative stress, ferroptosis, metabolic reprogramming, and pathways related to epigenetics and microbiota. This highlights the urgent need for regulatory measures and research on mechanisms to reduce developmental, reproductive, metabolic, and neurological risks tied to micro- and nanoplastics.

6. Conclusion

Nanoplastics are now detectable across tissues and fluids in multiple organisms, indicating that exposure leads to uptake, systemic distribution, and persistence at biologically relevant time scales. Once internalized, these particles interact with cellular membranes and macromolecules, form dynamic protein coronas, and localize to organelles such as mitochondria and lysosomes. The resulting cascade—membrane perturbation, oxidative stress, inflammation, impaired autophagy and lysosomal function, mitochondrial dysfunction, genotoxic and epigenetic alterations, and disrupted endocrine and immune signaling—provides converging mechanistic evidence that nanoplastics can compromise cellular homeostasis. Evidence of translocation across physiological barriers, including the gut epithelium, blood–brain barrier, and placenta, together with trophic transfer along food webs and potential microbiome dysbiosis, underscores that effects are not confined to single tissues or species but propagate across levels of biological organization.

At the same time, important uncertainties remain. Environmental exposures are complex mixtures in which particle size, shape, surface chemistry, aging, and sorbed co-contaminants jointly determine hazard, and many experimental studies still rely on pristine model particles and supra-environmental doses. Progress now depends on harmonized nomenclature, well-characterized reference materials, and exposure metrics that reflect particle number, mass, and surface area. Priorities include robust in vivo dose–response studies at environmentally realistic concentrations, longitudinal designs that capture chronic and developmental windows of susceptibility, and integrative "One Health" approaches that link human, animal, and ecosystem health. Taken together, the weight of current evidence supports the conclusion that nanoplastics can accumulate in biological systems and elicit adverse cellular and physiological effects; closing key data gaps will allow risk assessors and regulators to move from precaution toward evidence-based exposure limits and targeted mitigation strategies.

Abbreviations

General terms & types of plastics

MPs	Microplastics
NPs	Nanoplastics

PS	Polystyrene
PE	Polyethylene
PP	Polypropylene
PVC	Polyvinyl Chloride
PET	Polyethylene Terephthalate
PMMA	Polymethyl Methacrylate
PU	Polyurethane
CPE	Chlorinated Polyethylene
PA	Polyamide (e.g., Nylon)
PA-66/Nylon 66	Polyamide-66 (Nylon-66)
ABS	Acrylonitrile Butadiene Styrene
BR	Butadiene Rubber
EAA	Ethylene Acrylic Acid Copolymer
FR	Fluoro Rubber
EVA	Ethylene-Vinyl Acetate
PEVA	Polyethylene Vinyl Acetate
PE-co-PP	Polyethylene-co-Polypropylene (copolymer)
PC	Polycarbonate
POM	Polyoxymethylene
PTFE	Polytetrafluoroethylene
PER	Poly Epichlorohydrin Rubber

Analytical & detection methods

SEM	Scanning Electron Microscopy
TEM	Transmission Electron Microscopy
FTIR	Fourier-Transform Infrared Spectroscopy
μFTIR	Micro-Fourier Transform Infrared Spectroscopy
Raman	Raman Spectroscopy
SERS	Surface-Enhanced Raman Scattering
Py-GC-MS	Pyrolysis Gas Chromatography–Mass Spectrometry
GC/MS	Gas Chromatography–Mass Spectrometry
FFF-MALS	Field-Flow Fractionation with Multi-Angle Light Scattering
ICP-MS	Inductively Coupled Plasma Mass Spectrometry

Toxicological & oxidative stress markers

ROS	Reactive Oxygen Species
iROS	Intracellular Reactive Oxygen Species
HO-1	Heme Oxygenase 1
SOD1/SOD2	Superoxide Dismutase 1/2
GPX1	Glutathione Peroxidase 1
GSH	Glutathione
MDA	Malondialdehyde
ER stress	Endoplasmic Reticulum Stress

Pollutants & environmental chemicals

PAHs	Polycyclic Aromatic Hydrocarbons
POPs	Persistent Organic Pollutants

Cellular & molecular biology markers

ATP	Adenosine Triphosphate
LC3-II/LC3B	Microtubule-Associated Protein 1 Light Chain 3 (autophagy marker; lipidated form = II)
p62	Sequestosome-1 (autophagy adaptor protein)
BECN1	Beclin-1 (autophagy regulator)
BCL2	B-Cell Lymphoma 2 (anti-apoptotic regulator)
CASPASE-3/CASPASE-7	Cysteine-Aspartic Proteases (apoptosis executioner caspases)
HSP27	Heat Shock Protein 27
PAI-1	Plasminogen Activator Inhibitor 1

Inflammatory & immune markers

IL-6	Interleukin 6
IL-8	Interleukin 8
TNF-α	Tumor Necrosis Factor Alpha
B/T cells	B lymphocytes/T lymphocytes
GABA	Gamma-Aminobutyric Acid
hCG	Human Chorionic Gonadotropin
GD	Gestational Day
PND	Postnatal Day

Cellular pathways & signalling

EMT	Epithelial-Mesenchymal Transition
MMP2	Matrix Metalloproteinase 2
NOX4	NADPH Oxidase 4
MAPK/ERK	Mitogen-Activated Protein Kinase/Extracellular Signal-Regulated Kinase Pathway
PI3K/AKT	Phosphoinositide 3-Kinase/Protein Kinase B Pathway
MST1	Mammalian Sterile 20-like Kinase 1
LATS1	Large Tumor Suppressor Kinase 1
YAP1	Yes-Associated Protein 1
CTGF	Connective Tissue Growth Factor
CYR61	Cysteine-Rich Angiogenic Inducer 61
NOTCH3	Notch Receptor 3
TP53	Tumor Protein p53
GADD45A	Growth Arrest and DNA Damage-Inducible Alpha
RAD51	RAD51 Recombinase (DNA repair)

XRCC1	X-Ray Repair Cross-Complementing Protein 1
PPAR	Peroxisome Proliferator-Activated Receptor (metabolic regulator)
JAK-STAT	Janus Kinase/Signal Transducer and Activator of Transcription Pathway
MAPT	Microtubule-Associated Protein Tau
KIF21A	Kinesin Family Member 21A
Wnt	Wingless/Integrated Signalling Pathway
Shh	Sonic Hedgehog Pathway
NAFLD	Non-Alcoholic Fatty Liver Disease.

Author details

John Dennis and Tapan Kumar Mistri*
Department of Chemistry, SRM Institute of Science and Technology, Chennai, India

*Address all correspondence to: tapankum@srmist.edu.in

References

[1] Stanley J, Culliton D, Jovani-Sancho A-J, Neves AC. The journey of plastics: Historical development, environmental challenges, and the emergence of bioplastics for single-use products. English. 2025;**6**(1):17

[2] Babaremu KO, Okoya SA, Hughes E, Tijani B, Teidi D, Akpan A, Igwe J, Karera S, Oyinlola M, Akinlabi ET. Sustainable plastic waste management in a circular economy. Heliyon. 2022;**8** (7):e09984

[3] Dey S, Veerendra GTN, Babu PSSA, Manoj AVP, Nagarjuna K. Degradation of plastics waste and its effects on biological ecosystems: A scientific analysis and comprehensive review. Biomedical Materials & Devices. 2024;**2** (1):70–112

[4] Tsochatzis ED, Gika H, Theodoridis G, Maragou N, Thomaidis N, Corredig M. Microplastics and nanoplastics: Exposure and toxicological effects require important analysis considerations. Heliyon. 2024;**10**(11):e32261

[5] Preda O-T, Vlasceanu A-M, Andreescu CV, Tsatsakis A, Mezhuev Y, Negrei C, Baconi DL. Health implications of widespread micro- and nanoplastic exposure: Environmental prevalence, mechanisms, and biological impact on humans. Toxics. 2024;**12**(10):730

[6] de Oliveira CRS, Maestri G, Tochetto GA, de Oliveira JL, Stiegelmaier E, Fischer TV, Immich APS. Nanoplastics: Unveiling contamination routes and toxicological implications for human health. Current Analytical Chemistry. 2025a;**21**(3): 175–190

[7] Shukla S, Khanna S, Khanna K. Unveiling the toxicity of micro-nanoplastics: A systematic exploration of understanding environmental and health implications. Toxicology Reports. 2025;**14**:101844

[8] Wang R, Luan Y, Li J, Li X, Dai W, Tao K. Strong binding between nanoplastic and bacterial proteins facilitates protein corona formation and reduces nanoplastics toxicity. Science of the Total Environment. 2024;**951**:175433

[9] Yang H, Chen Z, Kong L, Xing H, Yang Q, Wu J. A review of eco-corona formation on micro/nanoplastics and its effects on stability, bioavailability, and toxicity. Water. 2025;**17**(8):1124

[10] Belioka M-P, Achilias D (2024). How natural phenomena and disasters together with the weathering conditions affect microplastics and nanoplastics in agricultural soils and in farmlands.

[11] Xu B, Zhao S, Li Y, Li J, Ma J, He K, Lu L, Wang X, Chen D, Li J. Elevated wet deposition of micro-and nano-plastics in remote mountains driven by free tropospheric transport. Journal of Hazardous Materials. 2025;**498**:140010

[12] Li M, Zhao Z, Zhao Z, Li M. Review of techniques for the detection, removal, and transformation of environmental microplastics and nanoplastics. ACS Applied Materials & Interfaces. 2025;**17**(14):20560–20589

[13] Rivera-Rivera DM, Quintanilla-Villanueva GE, Luna-Moreno D, Sánchez-Álvarez A, Rodríguez-Delgado JM, Cedillo-González EI, Kaushik G, Villarreal-Chiu JF, Rodríguez-Delgado

MM. Exploring innovative approaches for the analysis of micro- and nanoplastics: Breakthroughs in (bio) sensing techniques.. Biosensors. 2025;**15** (1):44

[14] De-la-torre GE, Bapp C, Forero-López AD, Dobaradaran S. Analytical techniques for quantifying and identifying nanoplastics: Recent advances. Current Opinion in Chemical Engineering. 2025;**48**:101134

[15] de Oliveira CRS, Maestri G, Tochetto GA, de Oliveira JL, Stiegelmaier E, Fischer TV, Immich APS. Nanoplastics: Unveiling contamination routes and toxicological implications for human health. Current Analytical Chemistry. 2025b;**21**(3):175–190

[16] Thakur R, Joshi V, Sahoo GC, Jindal N, Tiwari RR, Rana S. Review of mechanisms and impacts of nanoplastic toxicity in aquatic organisms and potential impacts on human health. Toxicology Reports. 2025;**102013**

[17] Dennis J, Arulraj D, Mistri TK. Unseen toxins: Exploring the human health consequences of micro and nanoplastics. Toxicology Reports. 2025;**14**:101955. DOI: 10.1016/j.toxrep.2025.101955

[18] da Silva Brito WA, Mutter F, Wende K, Cecchini AL, Schmidt A, Bekeschus S. Consequences of nano and microplastic exposure in rodent models: The known and unknown. Particle and Fibre Toxicology. 2022;**19**(1):28

[19] Shi Q, Tang J, Liu R, Wang L. Toxicity in vitro reveals potential impacts of microplastics and nanoplastics on human health: A review. Critical Reviews in Environmental Science and Technology. 2022;**52** (21):3863–3895

[20] Lu -Y-Y, Li H, Ren H, Zhang X, Huang F, Zhang D, Huang Q, Zhang X. Size-dependent effects of polystyrene nanoplastics on autophagy response in human umbilical vein endothelial cells. Journal of Hazardous Materials. 2022;**421**:126770

[21] Annangi B, Villacorta A, Vela L, Tavakolpournegari A, Marcos R, Hernández A. Effects of true-to-life PET nanoplastics using primary human nasal epithelial cells. Environmental Toxicology and Pharmacology. 2023;**100**: 104140

[22] Banerjee A, Billey LO, Shelver WL. Uptake and toxicity of polystyrene micro/nanoplastics in gastric cells: Effects of particle size and surface functionalization. PloS One. 2021;**16** (12):e0260803

[23] He Y, Li J, Chen J, Miao X, Li G, He Q, Xu H, Li H, Wei Y. Cytotoxic effects of polystyrene nanoplastics with different surface functionalization on human HepG2 cells. Science of the Total Environment. 2020;**723**:138180

[24] Annangi B, Villacorta A, López-Mesas M, Fuentes-Cebrian V, Marcos R, Hernández A. Hazard assessment of polystyrene nanoplastics in primary human nasal epithelial cells, focusing on the autophagic effects. Biomolecules. 2023;**13**(2):220

[25] Chen G, Xiong S, Jing Q, van Gestel CAM, van Straalen NM, Roelofs D, Sun L, Qiu H. Maternal exposure to polystyrene nanoparticles retarded fetal growth and triggered metabolic disorders of placenta and fetus in mice. Science of the Total Environment. 2023a;**854**:158666. DOI: 10.1016/j.scitotenv.2022.158666

[26] Mercer GV, Harvey NE, Steeves KL, Schneider CM, Sled JG, Macgowan CK, Baschat AA, Kingdom JC, Simpson AJ, Simpson MJ. Maternal exposure to polystyrene nanoplastics alters fetal brain metabolism in mice. Metabolomics. 2023;**19**(12):96

[27] Tamayo-Belda M, Pérez-Olivares AV, Pulido-Reyes G, Martin-Betancor K, González-Pleiter M, Leganés F, Mitrano D, Rosal R, Fernández-Piñas F. Tracking nanoplastics in freshwater microcosms and their impacts to aquatic organisms. Journal of Hazardous Materials. 2023;**445**:130625

[28] Nwankwo CEI, Okeke ES, Umeoguaju FU, Ejeromedoghene O, Adedipe DT, Ezeorba TPC. Addressing emerging contaminants in agriculture affecting plant–soil interaction: A review on bio-based and nano-enhanced strategies for soil health and global food security (GFS). Discover Toxicology. 2025;**2**(1):1–30

[29] Zhou P, Wang L, Gao J, Jiang Y, Adeel M, Hou D. Nanoplastic–plant interaction and implications for soil health. Soil Use and Management. 2023;**39**(1):13–42

[30] Bhagat J, Zang L, Nishimura N, Shimada Y. Application of omics approaches for assessing microplastic and nanoplastic toxicity in fish and seafood species. TrAC Trends in Analytical Chemistry. 2022;**154**:116674

[31] Rotchell JM, Jenner LC, Chapman E, Bennett RT, Bolanle IO, Loubani M, Sadofsky L, Hobkirk J, Palmer TM. Detection of microplastics in human saphenous vein tissue using μFTIR: A pilot study. PLoS One. 2023a;**18**(2):e0280594

[32] Ruan X, Ao J, Ma M, Jones RR, Liu J, Li K, Ge Q, Xu G, Liu Y, Wang T. Nanoplastics detected in commercial sea salt. Environmental Science & Technology. 2024;**58**(21):9091–9101

[33] Zhang J, Peng M, Lian E, Xia L, Asimakopoulos AG, Luo S, Wang L. Identification of poly(ethylene terephthalate) nanoplastics in commercially bottled drinking water using surface-enhanced Raman spectroscopy. Environmental Science & Technology. 2023;**57**(22):8365–8372

[34] Basini G, Bussolati S, Andriani L, Grolli S, Bertini S, Iemmi T, Menozzi A, Quintavalla F, Ramoni R, Serventi P. The effects of nanoplastics on adipose stromal cells from swine tissues. Domestic Animal Endocrinology. 2022;**81**:106747

[35] Martin-Folgar R, González-Caballero MC, Torres-Ruiz M, Cañas-Portilla AI, de Alba González M, Liste I, Morales M. Molecular effects of polystyrene nanoplastics on human neural stem cells. Plos One. 2024;**19**(1): e0295816

[36] Poinsignon L, Lefrère B, Azzouz AB, Chissey A, Colombel J, Djelidi R, Ferecatu I, Fournier T, Beaudeux J-L, Lespes G. Exposure of the human placental primary cells to nanoplastics induces cytotoxic effects, an inflammatory response and endocrine disruption. Journal of Hazardous Materials. 2025;**490**:137713

[37] Xiao S, Wang J, Digiacomo L, Amici A, De Lorenzi V, Pugliese LA, Cardarelli F, Cerrato A, Laganà A, Cui L. Protein corona alleviates adverse biological effects of nanoplastics in breast cancer cells. Nanoscale. 2024;**16**(35):16671–16683

[38] Yang Q, Zhang S, Su J, Li S, Lv X, Chen J, Lai Y, Zhan J. Identification of trace polystyrene nanoplastics down to 50 nm by the hyphenated method of filtration and surface-enhanced Raman spectroscopy based on silver nanowire membranes. Environmental Science & Technology. 2022a;**56**(15):10818–10828

[39] Yang T, Luo J, Nowack B. Characterization of nanoplastics, fibrils, and microplastics released during washing and abrasion of polyester textiles. Environmental Science & Technology. 2021;**55**(23):15873–15881

[40] Yee MS-L, Hii L-W, Looi CK, Lim W-M, Wong S-F, Kok -Y-Y, Tan B-K, Wong C-Y, Leong C-O. Impact of microplastics and nanoplastics on human health. Nanomaterials. 2021;**11**(2):496

[41] Stine JS, Aziere N, Harper BJ, Harper SL. A novel approach for identifying nanoplastics by assessing deformation behavior with scanning electron microscopy. Micromachines. 2023;**14**(10):1903

[42] Tanaka K, Takahashi Y, Kuramochi H, Osako M, Tanaka S, Suzuki G. Preparation of nanoscale particles of five major polymers as potential standards for the study of nanoplastics. Small. 2021;**17**(49):2105781

[43] Fang C, Luo Y, Naidu R. Raman imaging for the analysis of silicone microplastics and nanoplastics released from a kitchen sealant. Frontiers in Chemistry. 2023;**11**:1165523

[44] Hernandez LM, Yousefi N, Tufenkji N. Are there nanoplastics in your personal care products? Environmental Science & Technology Letters. 2017;**4**(7):280–285

[45] Du T, Yu X, Shao S, Li T, Xu S, Wu L. Aging of nanoplastics significantly affects protein corona composition thus enhancing macrophage uptake. Environmental Science & Technology. 2023;**57**(8):3206–3217

[46] Schiferle EB, Ge W, Reinhard BM. Nanoplastics weathering and polycyclic aromatic hydrocarbon mobilization. ACS Nano. 2023;**17**(6):5773–5784

[47] Xu H, Dong C, Yu Z, Hu Z, Yu J, Ma D, Yao W, Qi X, Ozaki Y, Xie Y. First identification of microplastics in human uterine fibroids and myometrium. Environmental Pollution. 2024;**360**:124632

[48] Fraissinet S, De Benedetto GE, Malitesta C, Holzinger R, Materić D. Microplastics and nanoplastics size distribution in farmed mussel tissues. Communications Earth & Environment. 2024;**5**(1):128

[49] Yang Q, Zhang S, Su J, Li S, Lv X, Chen J, Lai Y, Zhan J. Identification of trace polystyrene nanoplastics down to 50 nm by the hyphenated method of filtration and surface-enhanced Raman spectroscopy based on silver nanowire membranes. Environmental Science & Technology. 2022b;**56**(15):10818–10828

[50] Amobonye A, Bhagwat P, Raveendran S, Singh S, Pillai S. Environmental impacts of microplastics and nanoplastics: A current overview. Frontiers in Microbiology. 2021;**12**:768297

[51] Jiménez-Lamana J, Marigliano L, Allouche J, Grassl B, Szpunar J, Reynaud S. A novel strategy for the detection and quantification of nanoplastics by single particle inductively coupled plasma mass

spectrometry (ICP-MS). Analytical Chemistry. 2020;**92**(17):11664–11672

[52] Lee WS, Cho H-J, Kim E, Huh YH, Kim H-J, Kim B, Kang T, Lee J-S, Jeong J. Bioaccumulation of polystyrene nanoplastics and their effect on the toxicity of Au ions in zebrafish embryos. Nanoscale. 2019;**11**(7):3173–3185

[53] Schultz CL, Bart S, Lahive E, Spurgeon DJ. What is on the outside matters—surface charge and dissolve organic matter association affect the toxicity and physiological mode of action of polystyrene nanoplastics to C. elegans. Environmental Science & Technology. 2021;**55**(9):6065–6075

[54] Xiao F, Feng L-J, Sun X-D, Wang Y, Wang Z-W, Zhu F-P, Yuan X-Z. Do polystyrene nanoplastics have similar effects on duckweed (Lemna minor L.) at environmentally relevant and observed-effect concentrations? Environmental Science & Technology. 2022;**56**(7):4071–4079

[55] Molina S, Ocaña-Biedma H, Rodríguez-Sáez L, Landaburu-Aguirre J. Experimental evaluation of the process performance of MF and UF membranes for the removal of nanoplastics. Membranes. 2023;**13**(7):683

[56] Jiang W, Liu Y, Wu Y, Zhang L, Zhang B, Zhou S, Zhang P, Xu T, Wu M, Lv S. Polystyrene nanoplastics of different particle sizes regulate the polarization of pro-inflammatory macrophages. Scientific Reports. 2024;**14**(1):16329

[57] Larue C, Sarret G, Castillo-Michel H, Pradas Del Real AE. A critical review on the impacts of nanoplastics and microplastics on aquatic and terrestrial photosynthetic organisms. Small. 2021;**17**(20):2005834

[58] Cassano D, Bogni A, La Spina R, Gilliland D, Ponti J. Investigating the cellular uptake of model nanoplastics by single-cell ICP-MS. Nanomaterials. 2023;**13**(3):594

[59] Liang B, Zhong Y, Huang Y, Lin X, Liu J, Lin L, Hu M, Jiang J, Dai M, Wang B. Underestimated health risks: Polystyrene micro-and nanoplastics jointly induce intestinal barrier dysfunction by ROS-mediated epithelial cell apoptosis. Particle and Fibre Toxicology. 2021;**18**(1):20

[60] Dang F, Wang Q, Yan X, Zhang Y, Yan J, Zhong H, Zhou D, Luo Y, Zhu Y-G, Xing B. Threats to terrestrial plants from emerging nanoplastics. ACS Nano. 2022;**16**(10):17157–17167

[61] Roslan NS, Lee YY, Ibrahim YS, Anuar ST, Yusof KMKK, Lai LA, Brentnall T. Detection of microplastics in human tissues and organs: A scoping review. Journal of Global Health. 2024b;**14**:04179

[62] Rotchell JM, Jenner LC, Chapman E, Bennett RT, Bolanle IO, Loubani M, Sadofsky L, Hobkirk J, Palmer TM. Detection of microplastics in human saphenous vein tissue using μFTIR: A pilot study. PLoS One. 2023b;**18**(2):e0280594

[63] Amato-Lourenço LF, Dantas KC, Júnior GR, Paes VR, Ando RA, de Oliveira Freitas R, da Costa OMMM, Rabelo RS, Bispo KCS, Carvalho-Oliveira R. Microplastics in the olfactory bulb of the human brain. JAMA Network Open. 2024;**7**(9):e2440018–e2440018

[64] Nihart AJ, Garcia MA, El Hayek E, Liu R, Olewine M, Kingston JD, Castillo EF, Gullapalli RR, Howard T, Bleske B. Bioaccumulation of

microplastics in decedent human brains. Nature Medicine. 2025;**31**(4):1114–1119

[65] Marfella R, Prattichizzo F, Sardu C, Fulgenzi G, Graciotti L, Spadoni T, D'Onofrio N, Scisciola L, La Grotta R, Frigé C. Microplastics and nanoplastics in atheromas and cardiovascular events. New England Journal of Medicine. 2024;**390**(10):900–910

[66] La Porta E, Exacoustos O, Lugani F, Angeletti A, Chiarenza DS, Bigatti C, Spinelli S, Kajana X, Garbarino A, Bruschi M. Microplastics and kidneys: An update on the evidence for deposition of plastic microparticles in human organs, tissues and fluids and renal toxicity concern. International Journal of Molecular Sciences. 2023;**24**(18):14391

[67] Huang S, Huang X, Bi R, Guo Q, Yu X, Zeng Q, Huang Z, Liu T, Wu H, Chen Y. Detection and analysis of microplastics in human sputum. Environmental Science & Technology. 2022;**56**(4):2476–2486

[68] Sun A, Wang W-X. Human exposure to microplastics and its associated health risks. Environment & Health. 2023;**1**(3):139–149

[69] Lai H, Liu X, Qu M. Nanoplastics and human health: Hazard identification and biointerface. Nanomaterials. 2022;**12**(8):1298

[70] Prata JC. Microplastics and human health: Integrating pharmacokinetics. Critical Reviews in Environmental Science and Technology. 2023;**53**(16):1489–1511

[71] Pironti C, Notarstefano V, Ricciardi M, Motta O, Giorgini E, Montano L. 5. First evidence of microplastics in human urine, a preliminary study of intake in the human body. Toxics. 2022a;**11**(1):40

[72] Roslan NS, Lee YY, Ibrahim YS, Anuar ST, Yusof KMKK, Lai LA, Brentnall T. Detection of microplastics in human tissues and organs: A scoping review. Journal of Global Health. 2024a;**14**:04179

[73] Ragusa A, Notarstefano V, Svelato A, Belloni A, Gioacchini G, Blondeel C, Zucchelli E, De Luca C, D'Avino S, Gulotta A. Raman microspectroscopy detection and characterisation of microplastics in human breastmilk. Polymers. 2022a;**14**(13):2700

[74] Ragusa A, Notarstefano V, Svelato A, Belloni A, Gioacchini G, Blondeel C, Zucchelli E, De Luca C, D'Avino S, Gulotta A. Raman microspectroscopy detection and characterisation of microplastics in human breastmilk. Polymers. 2022b;**14**(13):2700

[75] Pironti C, Notarstefano V, Ricciardi M, Motta O, Giorgini E, Montano L. First evidence of microplastics in human urine, a preliminary study of intake in the human body. Toxics. 2022b;**11**(1):40

[76] Montano L, Giorgini E, Notarstefano V, Notari T, Ricciardi M, Piscopo M, Motta O. Raman microspectroscopy evidence of microplastics in human semen. Science of the Total Environment. 2023;**901**:165922

[77] Zhao T, Shen L, Ye X, Bai G, Liao C, Chen Z, Peng T, Li X, Kang X, An G. Prenatal and postnatal exposure to polystyrene microplastics induces testis developmental disorder and affects

male fertility in mice. Journal of Hazardous Materials. 2023;**445**:130544

[78] Zhong Y, Yang Y, Zhang L, Ma D, Wen K, Cai J, Cai Z, Wang C, Chai X, Zhong J. Revealing new insights: Two-center evidence of microplastics in human vitreous humor and their implications for ocular health. Science of the Total Environment. 2024;**921**:171109

[79] Guo X, Wang L, Wang X, Li D, Wang H, Xu H, Liu Y, Kang R, Chen Q, Zheng L. Discovery and analysis of microplastics in human bone marrow. Journal of Hazardous Materials. 2024;**477**:135266

[80] Sun J, Sui M, Wang T, Teng X, Sun J, Chen M. Detection and quantification of various microplastics in human endometrium based on laser direct infrared spectroscopy. Science of the Total Environment. 2024;**906**:167760

[81] Codrington J, Varnum AA, Hildebrandt L, Pröfrock D, Bidhan J, Khodamoradi K, Höhme A-L, Held M, Evans A, Velasquez D. Detection of microplastics in the human penis. International Journal of Impotence Research. 2024;1–7

[82] Cortés C, Domenech J, Salazar M, Pastor S, Marcos R, Hernández A. Nanoplastics as a potential environmental health factor: Effects of polystyrene nanoparticles on human intestinal epithelial Caco-2 cells. Environmental Science: Nano. 2020;**7**(1):272–285

[83] Domenech J, de Britto M, Velázquez A, Pastor S, Hernández A, Marcos R, Cortés C. Long-term effects of polystyrene nanoplastics in human intestinal Caco-2 cells. Biomolecules. 2021;**11**(10):1442

[84] Li Y, Li Y, Li J, Song Z, Zhang C, Guan B. Toxicity of polystyrene nanoplastics to human embryonic kidney cells and human normal liver cells: Effect of particle size and Pb2+ enrichment. Chemosphere. 2023;**328**:138545

[85] Halimu G, Zhang Q, Liu L, Zhang Z, Wang X, Gu W, Zhang B, Dai Y, Zhang H, Zhang C. Toxic effects of nanoplastics with different sizes and surface charges on epithelial-to-mesenchymal transition in A549 cells and the potential toxicological mechanism. Journal of Hazardous Materials. 2022;**430**:128485

[86] Zeng L, Zhou C, Xu W, Huang Y, Wang W, Ma Z, Huang J, Li J, Hu L, Xue Y. The ovarian-related effects of polystyrene nanoplastics on human ovarian granulosa cells and female mice. Ecotoxicology and Environmental Safety. 2023;**257**:114941

[87] Binder ARD, Mussack V, Kirchner B, Pfaffl MW. Uptake and effects of polystyrene nanoplastics in comparison to non-plastic silica nanoparticles on small intestine cells (IPEC-J2). Ecotoxicology and Environmental Safety. 2023;**262**:115147

[88] Peng M, Vercauteren M, Grootaert C, Catarino AI, Everaert G, Rajkovic A, Janssen C, Asselman J. Bioenergetic effects of pristine and ultraviolet-weathered polydisperse polyethylene terephthalate and polystyrene nanoplastics on human intestinal Caco-2 cells. Science of the Total Environment. 2024;**908**:168267

[89] Banerjee A, Billey LO, McGarvey AM, Shelver WL. Effects of polystyrene micro/nanoplastics on liver cells based on particle size, surface functionalization, concentration and

exposure period. Science of the Total Environment. 2022;**836**:155621

[90] Fournier SB, D'Errico JN, Adler DS, Kollontzi S, Goedken MJ, Fabris L, Yurkow EJ, Stapleton PA. Nanopolystyrene translocation and fetal deposition after acute lung exposure during late-stage pregnancy. Particle and Fibre Toxicology. 2020;**17**(1):55

[91] Cary CM, Fournier SB, Adams S, Wang X, Yurkow EJ, Stapleton PA. Single pulmonary nanopolystyrene exposure in late-stage pregnancy dysregulates maternal and fetal cardiovascular function. Toxicological Sciences. 2024a;**199**(1):149–159

[92] Chen G, Xiong S, Jing Q, van Gestel CAM, van Straalen NM, Roelofs D, Sun L, Qiu H. Maternal exposure to polystyrene nanoparticles retarded fetal growth and triggered metabolic disorders of placenta and fetus in mice. Science of the Total Environment. 2023b;**854**:158666

[93] Dibbon KC, Mercer GV, Maekawa AS, Hanrahan J, Steeves KL, Ringer LCM, Simpson AJ, Simpson MJ, Baschat AA, Kingdom JC. Polystyrene micro-and nanoplastics cause placental dysfunction in mice. Biology of Reproduction. 2024;**110**(1):211–218

[94] Wang X, Zhao Z, Wang X, Hu W, Chu X, Qian M, Wang R, Yu S, Wu Q, Tang J. Effects of polystyrene nanoplastic gestational exposure on mice. Chemosphere. 2023;**324**:138255

[95] Luo T, Zhang Y, Wang C, Wang X, Zhou J, Shen M, Zhao Y, Fu Z, Jin Y. Maternal exposure to different sizes of polystyrene microplastics during gestation causes metabolic disorders in their offspring. Environmental Pollution. 2019;**255**:113122

[96] Jeong B, Kim J-S, Kwon AR, Lee J, Park S, Koo J, Lee WS, Baek JY, Shin W-H, Lee J-S. Maternal nanoplastic ingestion induces an increase in offspring body weight through altered lipid species and microbiota. Environment International. 2024;**185**:108522

[97] Dou Y, Zhang M, Zhang H, Zhang C, Feng L, Hu J, Gao Y, Yuan X-Z, Zhao Y, Zhao H. Lactating exposure to microplastics at the dose of infants ingested during artificial feeding induced reproductive toxicity in female mice and their offspring. Science of the Total Environment. 2024;**949**:174972

[98] Sun J, Teng M, Zhu W, Zhao X, Zhao L, Li Y, Zhang Z, Liu Y, Bi S, Wu F. MicroRNA and gut microbiota alter intergenerational effects of paternal exposure to polyethylene nanoplastics. ACS Nano. 2024;**18**(27):18085–18100

[99] Zhang Q, Lang Y, Tang X, Cheng W, Cheng Z, Rizwan M, Xie L, Liu Y, Xu H, Liu Y. Polystyrene microplastic-induced endoplasmic reticulum stress contributes to growth plate endochondral ossification disorder in young rat. Environmental Toxicology. 2024;**39**(6):3314–3329

[100] Chen J, Zhang Y, Liu X, Li K, Liu H, Lai W, Shi Y, Xi Z, Yan L, Tian L. Effects of exposure to nano-plastic drinking during pregnancy on cognitive related proteins in offspring of SD rats. Environmental Pollutants and Bioavailability. 2024;**36**(1):2292104

[101] Tian L, Zhang Y, Chen J, Liu X, Nie H, Li K, Liu H, Lai W, Shi Y, Xi Z. Effects of nanoplastic exposure during pregnancy and lactation on neurodevelopment of rat offspring. Journal of Hazardous Materials. 2024;**474**:134800

[102] Shin HS, Lee SH, Moon HJ, So YH, Lee HR, Lee E-H, Jung E-M. Exposure to polystyrene particles causes anxiety-, depression-like behavior and abnormal social behavior in mice. Journal of Hazardous Materials. 2023;**454**:131465

[103] Park E-J, Han J-S, Park E-J, Seong E, Lee G-H, Kim D-W, Son H-Y, Han H-Y, Lee B-S. Repeated-oral dose toxicity of polyethylene microplastics and the possible implications on reproduction and development of the next generation. Toxicology Letters. 2020;**324**:75–85

[104] Shang Q, Wu H, Wang K, Zhang M, Dou Y, Jiang X, Zhao Y, Zhao H, Chen Z-J, Wang J. Exposure to polystyrene microplastics during lactational period alters immune status in both male mice and their offspring. Science of the Total Environment. 2024;**951**:175371

[105] Tang J, Bu W, Hu W, Zhao Z, Liu L, Luo C, Wang R, Fan S, Yu S, Wu Q. Ferroptosis is involved in sex-specific small intestinal toxicity in the offspring of adult mice exposed to polystyrene nanoplastics during pregnancy. ACS Nano. 2023;**17**(3):2440–2449

[106] Cary CM, Fournier SB, Adams S, Wang X, Yurkow EJ, Stapleton PA. Single pulmonary nanopolystyrene exposure in late-stage pregnancy dysregulates maternal and fetal cardiovascular function. Toxicological Sciences. 2024b;**199**(1):149–159

Chapter 3

The Hidden Poison: How Microplastics and Nanoplastics Threaten the Health of Aquatic Organisms Across Ecosystems

Naveen Kumar, Jayendra Kumar Singh, Mrinmoy Mandal, Nisha Yadav and Divya Tyagi

Abstract

Synthetic polymer debris, particularly microplastics and nanoplastics (NPs), has emerged as a structurally persistent and biologically active stressor in aquatic systems. This chapter offers an integrative synthesis of how micro-/nanoplastics interact with aquatic organisms at molecular, cellular, and systemic levels. Beyond trophic ingestion, we examine sublethal toxicodynamics, including mitochondrial collapse, oxidative stress, neurotoxicity, immunomodulation, and reproductive disruption – effects now documented across phylogenetically diverse taxa. NPs, due to their ultrafine scale and surface reactivity, exhibit disproportionate bioavailability and organotropism, making them uniquely invasive at the intracellular level. We link mechanistic laboratory evidence with field-scale ecological data to trace bioaccumulation patterns and cascading disruptions across food webs. These particles are termed "hidden poison" by our laboratory in a recent paper because their effects are often cryptic, chronic, and cumulative – rarely lethal in isolation, yet profoundly destabilizing over time. Lacking scent, taste, or visibility, they bypass organismal defenses and regulatory frameworks alike. This silent toxicity challenges conventional paradigms of aquatic health assessment. By bridging ecotoxicology with systems biology and material science, this chapter proposes a new framework for understanding polymer-driven biological stress. We conclude with a call for predictive models that integrate particle-specific traits with ecosystem-level feedback, essential for mitigating this unresolved planetary contaminant.

Keywords: aquatic ecotoxicology, nanoplastics-induced cellular stress, trophic bioaccumulation, hidden pollutants, polymer-driven biological disruption

1. Introduction

1.1 Background on synthetic polymer pollution

The global surge in plastic production – from approximately 2 million tonnes in 1950 to over 460 million tonnes in 2019 – has introduced an unprecedented flux of synthetic polymers into aquatic ecosystems [1]. Their intrinsic resistance to abiotic and biotic degradation arises from high molecular weight, hydrophobicity, crystallinity, and stable C–C and C–H bonds, resulting in persistence over timescales that exceed ecological resilience thresholds [2].

Aquatic systems receive these polymers through multiple concurrent pathways (**Figure 1**). Major transport vectors include wastewater effluents, urban stormwater runoff, atmospheric fallout, and riverine inputs, with hydrodynamic forces determining eventual partitioning into pelagic, benthic, or estuarine compartments [3]. A high-resolution mass-balance survey of sedimentary MPs in Italy, applying the Polymer Identification and Specific Analysis protocol, revealed accumulation hot spots of oxidized polyolefins and PS (up to 1138 μg/kg sediment). These hot spots exhibited polymer-specific degradation products alongside cosorbed phthalates and brominated flame retardants [4]. Such particles were frequently enriched in hydrophobic organic contaminants (e.g., polycyclic aromatic hydrocarbons (PAHs), PCBs) and heavy metals, demonstrating their dual role as chemical vectors and environmental sinks.

Plastic pollution was historically framed as a physical hazard – manifested as entanglement and ingestion of macroplastics by marine vertebrates – but recent evidence indicates a chemical and biological threat. Submicron plastics (<1000 nm) exhibit high specific surface area and reactive aging surfaces, allowing epithelial interactions, bypass of endosomal entrapment, and activation of intracellular stress pathways [5]. In *Daphnia magna*, industrially relevant exposure induced ROS overproduction, mitochondrial dysfunction, and altered reproduction – outcomes that were sublethal yet ecologically disruptive.

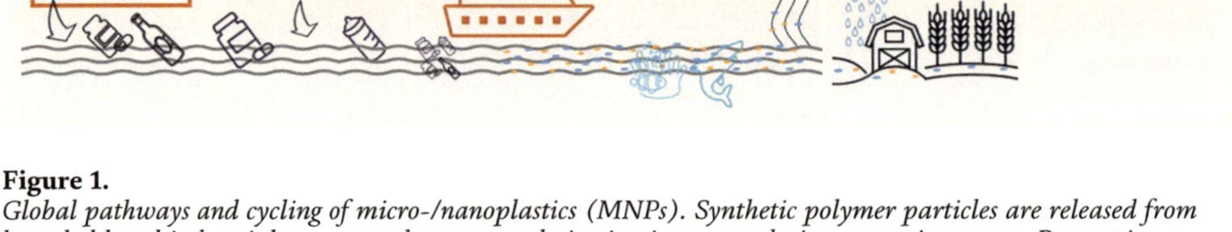

Figure 1.
Global pathways and cycling of micro-/nanoplastics (MNPs). Synthetic polymer particles are released from household and industrial sources and transported via riverine networks into aquatic systems. Progressive fragmentation drives the formation of secondary microplastics (MPs), which enter food webs through bioaccumulation in aquatic organisms. Volatilization and atmospheric transfer subsequently redistribute these particles into clouds and rainfall, enabling long-range transport and deposition into remote ecosystems, including croplands and polar regions.

Synthetic polymer distribution has expanded beyond coastlines to polar and alpine environments (**Figure 1**). Cryospheric snow from the Kola Peninsula contained synthetic polymer fibers, likely derived from long-range Aeolian transport and secondary atmospheric fragmentation [6]. This finding redefines aquatic contamination to include vertical atmospheric deposition as a critical dispersal mechanism.

Importantly, the ecological risk of MNPs has been systematically underestimated due to analytical limitations, regulatory blind spots, and the misconception of plastics as chemically inert. Their surface-mediated bioavailability, long-range mobility, and capacity for biological translocation establish them as mobile, bioactive pollutants capable of inducing multiscale ecological perturbations [7]. That is why our laboratory has proposed the term "hidden poison" to describe these particles [8]. This designation is justified because MNPs exert cryptic, persistent, and often cumulative effects that may not manifest immediately or through overt toxicity. Instead, their lack of scent, taste, and visibility allows them to bypass both biological and regulatory detection, silently destabilizing aquatic systems through chronic exposures that undermine organismal health and ecological resilience over time.

1.2 Definitions and scope: MPs vs. NPs

MNPs are size-defined fragments of synthetic polymers, yet their classification carries toxicological and environmental implications beyond dimensionality. MPs are typically defined as 1 μm–5 mm, whereas NPs are < 1 μm, with some frameworks restricting NPs to < 100 nm to reflect colloidal behavior, quantum surface phenomena, and rapid diffusivity [9, 10]. Regulatory definitions remain non-harmonized across ECHA, ISO, and OECD, complicating hazard assessment and inter-study comparability. Plastic particles are further bifurcated by origin into primary MNPs – engineered at micro-/nanoscales (e.g., cosmetic microbeads, nanoparticle additives in paints, nanocarriers in drug delivery systems) – and secondary MNPs – are generated from macroplastic degradation via photolysis (Norrish I/II), oxidative chain scission, mechanical abrasion, and hydrolysis, leading to progressively smaller, more oxidized, and morphologically irregular particles [11, 12].

Environmental persistence also diverges. MPs accumulate in benthic sediments and ocean gyres, persisting for decades, whereas NPs exhibit high inter-compartment mobility, being detected in cryospheric snow, atmospheric fallout, and intracellular matrices, supporting their role in long-range transport and chronic bioavailability [12, 13]. A major challenge is the absence of standardized NP metrology. MPs are detectable via μ-FTIR and Raman spectroscopy (>300 μm), while NPs require ultrafiltration, AF4, pyro-GC/MS, and DLS – methods rarely deployed in routine ecotoxicology [13]. Without harmonized detection frameworks, size transitions from a metric to a blind spot.

Building on this challenge, the chapter develops a vertically integrated framework linking MNPs to aquatic biological interactions across molecular, physiological, and ecosystem scales. Section 2 examines environmental sources and transformation processes; Sections 3 and 4 address uptake routes and toxicodynamic mechanisms; Section 5 explores physiological endpoints in aquatic organisms;

Section 6 considers trophic bioaccumulation and ecological feedback loops; and Section 7 outlines predictive modeling, detection strategies, and policy integration. By bridging particle-level properties with population- and ecosystem-level perturbations, the chapter provides a mechanistic lens for understanding the hidden toxicity of MNPs.

2. Environmental sources and physicochemical fate

2.1 Primary and secondary sources of micro-/nanoparticles

MNPs infiltrate aquatic environments through two principal pathways: primary emissions and secondary generation. This distinction is critical because secondary particles dominate environmental burdens, comprising over 90% of marine plastic fragments [14].

Primary MPs and NPs are intentionally engineered for industrial or commercial purposes. Historically, polyethylene (PE) and polypropylene (PP) microbeads were incorporated into cosmetics and personal care products (PCPs) (**Figure 2**). Despite regulatory bans, such as the US Microbead-Free Waters Act, legacy residues persist downstream of wastewater treatment plants, embedded in sediments [15]. Synthetic textiles represent another dominant primary source; polyester and nylon garments shed up to 700,000 fibers per wash, contributing to long-term retention in aquatic sediments [16]. Tire wear particles, generated from styrene-butadiene rubber, constitute the largest urban polymer input, releasing ~ 1.5 million metric tonnes annually into road runoff and air, and dominating stormwater polymer loads [17].

Secondary MPs and NPs originate from the fragmentation of macroplastic debris under photodegradation, oxidative weathering, and mechanical abrasion. For example, PE films and PET bottles rapidly fragment under UV irradiation and turbulence, with thin plastic bags disintegrating into MPs within 2 months of sunlight exposure and subsequently generating NPs under shear forces (**Figure 3**) [18]. Agricultural LDPE mulch films are similarly susceptible to oxidative cracking and plow-induced degradation, introducing persistent plastic fragments into soils and hydrologically connected waterways [16].

Atmospheric fallout is emerging as an underestimated vector for MNPs. Fibers from synthetic textiles, tire-derived particles, and urban dust have been detected in remote alpine air, arctic snow, and marine rainfall, with urban deposition rates reaching ~ 300 particles/m^2/day [17]. Their low density and colloidal behavior enable long-range atmospheric transport, effectively connecting terrestrial sources to aquatic sinks. Globally, over 11 million metric tonnes of plastic enter aquatic systems annually, a figure projected to escalate to 29 million metric tonnes by 2040 under current trends [16, 19].

2.2 Surface properties, aging, and toxic potential

The toxicological impact of MNPs arises not solely from polymer type but from dynamic alterations in surface chemistry and morphology during environmental aging. Virgin polymers – including PE, PP, and PS – are highly hydrophobic with low

DOI: http://dx.doi.org/10.5772/intechopen.1013156

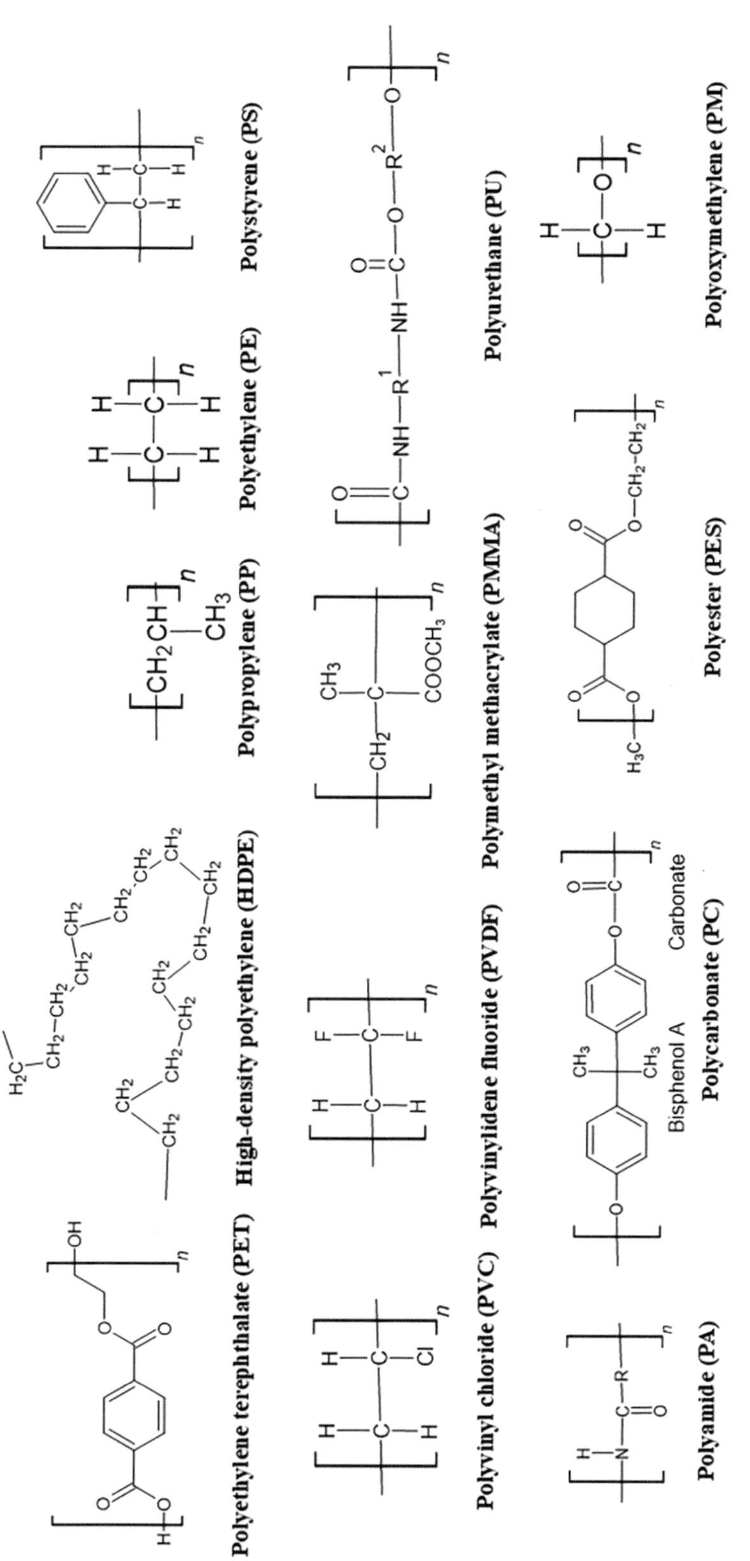

Figure 2.
Major synthetic polymers contributing to MP pollution. Representative chemical structures of widely used plastics are shown, which contribute to MP pollution, including polyethylene terephthalate (PET), high-density polyethylene, polypropylene (PP), polyethylene (PE), polystyrene (PS), polyvinyl chloride (PVC), polyvinylidene fluoride (PVDF), polymethyl methacrylate (PMMA), polyurethane (PU), polyamide (PA), polycarbonate (PC), polyester (PES), and polyoxymethylene (PM).

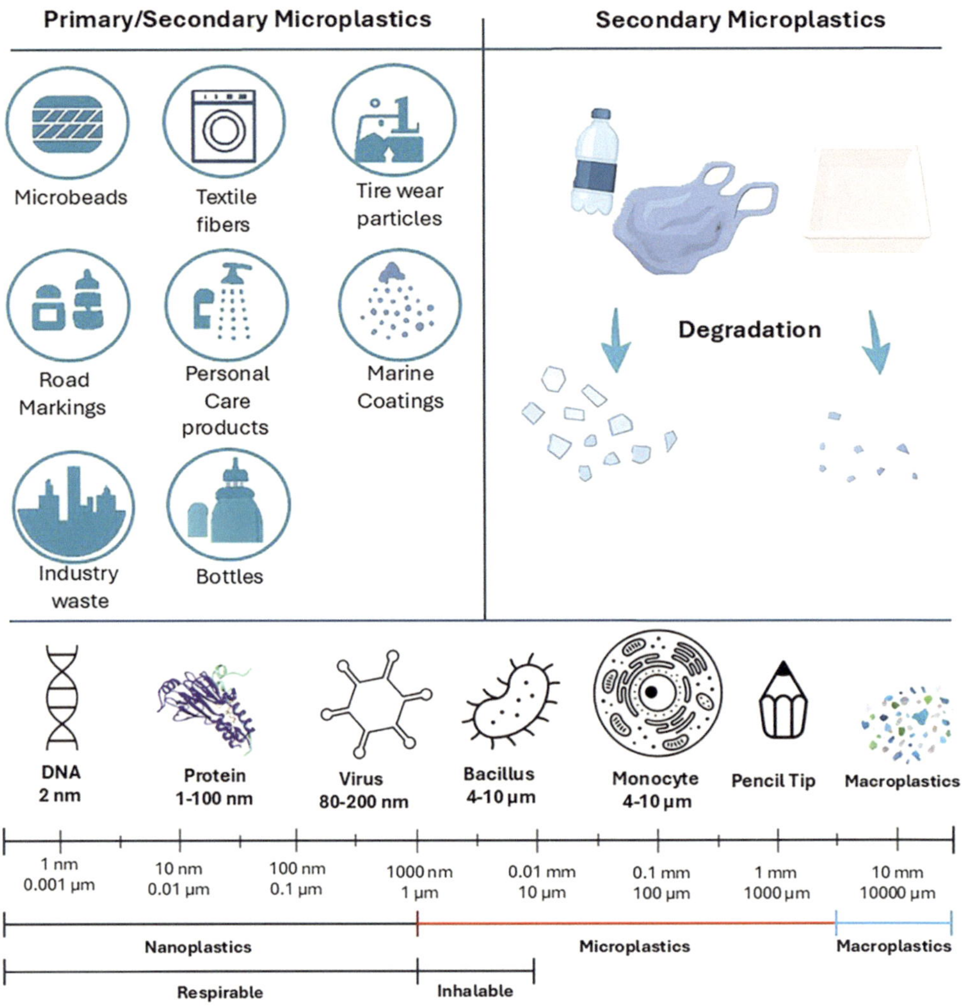

Figure 3.
Sources, fragmentation pathways, and size spectrum of primary and secondary MNPs. Primary MNPs originate from intentionally manufactured sources such as microbeads, textile fibers, tire wear particles, PCPs, marine coatings, road markings, bottles, and industrial waste. Secondary MNPs arise through the degradation and fragmentation of larger plastic items, including packaging films, bottles, and trays. The lower panel illustrates the environmental size continuum, comparing NPs (1–1000 nm) and MPs (1 µm–5 mm) to biological and particulate references (DNA, proteins, viruses, bacteria, monocytes, and macroscopic objects).

surface charge, facilitating passive diffusion across lipid membranes and bioadhesion to gill, gut, and hepatopancreatic tissues (**Figure 4**) [20]. Nanoscale dimensions (<1000 nm) amplify surface area-to-volume ratios, increasing pollutant sorption capacity and cellular interface contact. For example, 80-nm PS NPs were internalized into zebrafish (*Danio rerio*) hepatocytes within 4 hours, inducing dose-dependent ROS elevation and mitochondrial swelling at concentrations as low as 50 µg/L [21].

Environmental aging transforms particle surfaces in three principal ways:

DOI: http://dx.doi.org/10.5772/intechopen.1013156

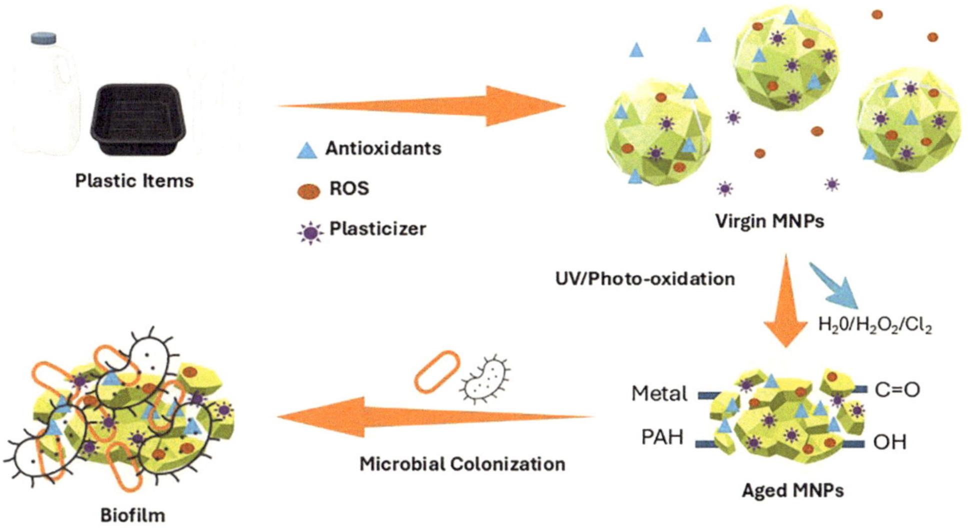

Figure 4.
Environmental aging and transformation of MNPs. Virgin MNPs, originating from consumer plastic items, contain embedded additives such as antioxidants and plasticizers. Upon exposure to UV light, oxidants (H_2O_2, Cl_2), and reactive oxygen species (ROS), these particles undergo surface oxidation, developing functional groups (C=O, –OH) and enhanced affinity for persistent organic pollutants (POPs) and metals. Subsequent microbial colonization and biofilm formation further modify surface chemistry, increasing particle reactivity, persistence, and biological uptake.

1. **Photooxidation** – UV exposure introduces hydroxyl (–OH), carbonyl (C=O), and carboxyl (–COOH) groups, increasing hydrophilicity and negative zeta potential. Aged PS particles displayed a 25% rise in the carbonyl index after 96 h, enhancing the adsorption of benzo[a]pyrene and Cd^{2+} from seawater [21].

2. **Mechanical erosion** – Abrasion produces irregular edges and pore structures, promoting biofilm colonization. Biofilm-coated MNPs demonstrated 4 × higher Pb^{2+} sorption and triggered immune activation in *Mytilus galloprovincialis* hemocytes [22].

3. **Leaching of embedded additives** – PVC fragments exposed to sunlight for 72 hours released BPA and phthalates at 3–5 × higher rates than unaged particles, activating estrogenic genes (VTG1, ERα) in medaka liver cells [23]. Flame retardants and UV stabilizers, once leached, synergize with sorbed contaminants, amplifying endocrine and immunotoxic effects.

Surface aging also modifies colloidal dynamics and bioavailability. Negatively charged NPs exhibit enhanced dispersion and cellular uptake [23]. Ultimately, aged MNPs function as multiphase toxic vectors, capable of binding, concentrating, and delivering POPs, metal ions, and microbial toxins to intracellular compartments. They impose mechanical stress via cellular penetration and biochemical disruption through cotransported toxics – a process further elaborated in Section 3 on uptake and exposure pathways.

2.3 Fate in aquatic environments: Aggregation and transport

After entry into aquatic systems, the fate of MNPs is dictated by interfacial interactions, hydrodynamic forces, and geochemical conditions. Aggregation – both homoaggregation (self-association) and heteroaggregation (binding with natural colloids or biogenic particles) – is the primary driver of vertical flux, colloidal stability, and sediment sequestration (**Figure 5**).

Heteroaggregation dominates in natural waters. Drivers include particle size, surface charge (zeta potential), oxidation state, and bridging by divalent cations or biopolymers. Sedimentary organic matter, particularly amorphous and black carbon, facilitates aggregation via hydrogen bonding and reduced electrostatic repulsion, with salinity-dependent variations [24].

Ionic conditions strongly modulate aggregation. Ca^{2+} and Mg^{2+} ions compress electrical double layers and induce bridging flocculation, as predicted by DLVO/XDLVO theory. In contrast, macromolecules such as EPS, BSA, and humic acids provide electrosteric stabilization at low ionic strength but enhance aggregation under divalent-rich conditions by collapsing hydration barriers [25, 26]. Surface interactions with nanomaterials further influence fate: MoS_2 nanosheets stabilize MPs in suspension, whereas graphene oxide promotes macroaggregate formation via van der Waals adhesion and oxidative edge interactions [27].

Vertical transport depends on density, aggregation state, and biofouling history. Low-density polymers (PE, PP) initially float, but biofilm growth and detritus adsorption increase effective density, promoting sinking. NPs often remain suspended due to Brownian motion and electrostatic stabilization. Seventy-nanometer PS NPs, for example, exhibit critical coagulation concentrations far exceeding typical freshwater ion levels, preventing aggregation unless destabilized by pH or ionic shifts [28]. In estuarine waters, however, elevated salinity and temperature drive rapid aggregation, forming > 2 μm clusters that accelerate sedimentation [29].

Benthic sediments act as both sinks and conditional sources. Sediment-bound MNPs are influenced by deposition rates, organic matter content, and redox-driven transformations. Though persistent over decadal scales, they undergo biofilm colonization and additive leaching, which can release pollutants into porewater, posing benthic ecotoxicological risks [30]. Bioturbation by infaunal organisms redistributes buried particles, while freeze–thaw cycles in cold systems immobilize NPs in cryogenic strata [31].

Horizontal transport occurs via fluvial discharge, estuarine mixing, coastal currents, and storm-induced resuspension. High-energy events – such as storms or dredging – can exceed critical shear stress, resuspending previously sequestered plastics. Process-based models, such as Full Multi Transport, demonstrate that particle speciation (pristine, biofouled, heteroaggregated) and aggregation state critically determine lateral advection across stratified waters [32].

During transport, continuous physicochemical aging – including photooxidation and biofouling – modifies surface charge, hydrophilicity, and contaminant affinity. Oxygenated surfaces display enhanced sorption for trace metals and hydrophobic organic pollutants, transforming MNPs into mobile, reactive vectors capable of cross-compartment contaminant transport [33, 34]. This dynamic fate underpins their ecotoxicological complexity in aquatic environments.

DOI: http://dx.doi.org/10.5772/intechopen.1013156

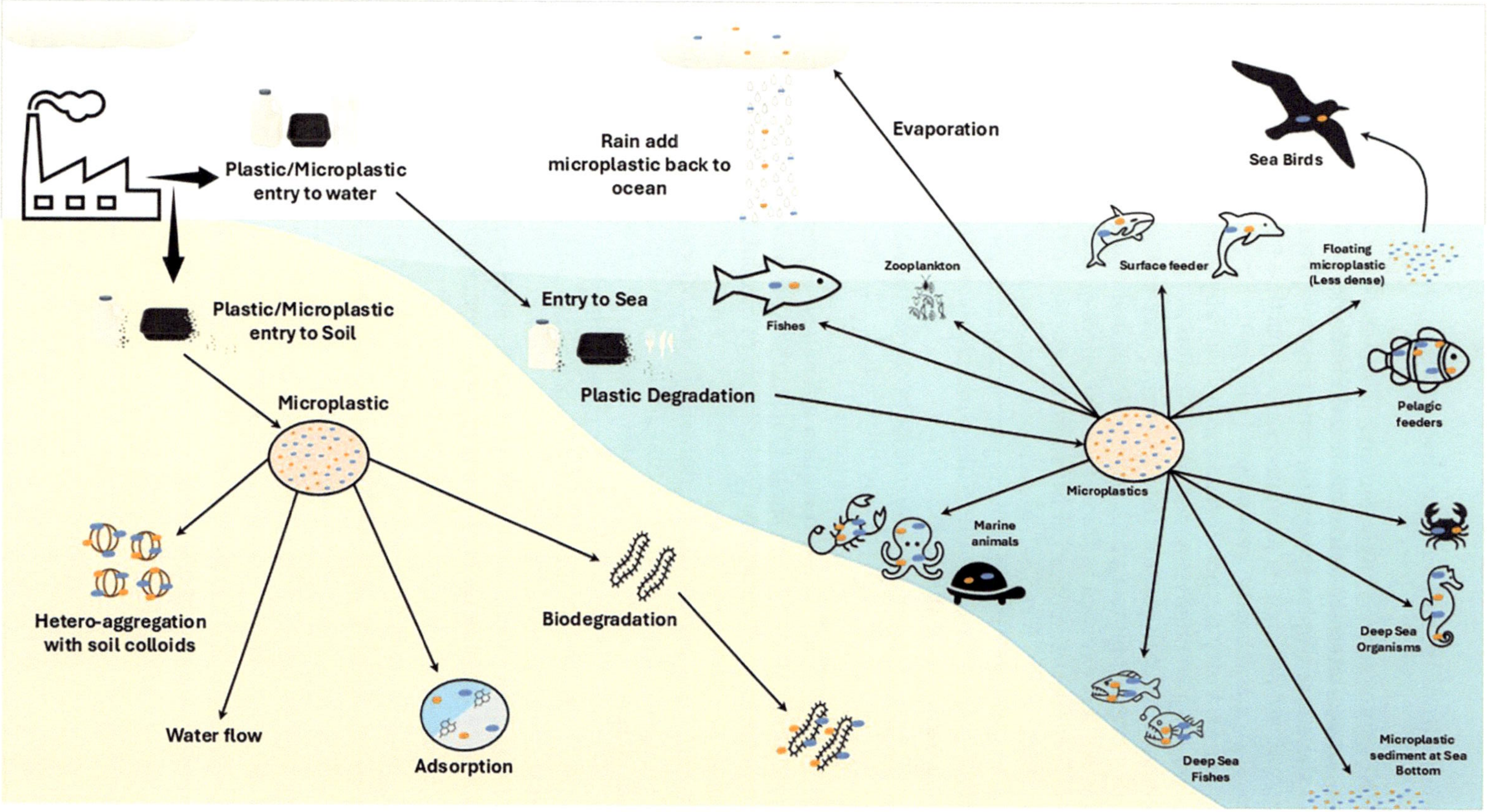

Figure 5.
Environmental cycling, degradation, and trophic transfer of MPs. Plastics and MPs enter aquatic and terrestrial systems through industrial and household discharges, subsequently partitioning into soil and water. In soils, particles undergo hetero-aggregation with colloids, adsorption, and water-driven transport, while in aquatic systems, they degrade and disperse through surface waters, sediments, and biota. Microplastics are ingested across trophic levels – from zooplankton and surface-feeding fish to pelagic and deep-sea organisms – and accumulate in marine animals and seabirds. Floating fractions remain buoyant, whereas denser particles settle into sediments, creating long-term benthic reservoirs. Evaporation and rainfall recycle particles into atmospheric and oceanic pathways, underscoring their persistence and cross-ecosystem mobility.

3. Exposure pathways and organismal entry

3.1 Ingestion, filtration, and surface absorption

Once MNPs enter aquatic environments, they exploit evolutionarily conserved feeding and surface-interface behaviors to gain biological entry. These exposure routes span the entire trophic spectrum, from phytoplankton and rotifers to apex vertebrates, and are dictated by a convergence of particle traits (size, charge, hydrophobicity), anatomical access points (gills, gut, skin, mucosa), and species-specific behaviors. Entry is, therefore, a mechanistically precise process, not a random consequence of environmental availability.

Ingestion initiates at the base of the food web. Filter-feeding protists, rotifers, and ciliates readily engulf polymer particles 0.5–10 μm in diameter (**Figure 6**). Remarkably, rotifers exposed to fluorescent PS microspheres demonstrated intracellular fragmentation of larger MPs into nanoscale particles, suggesting in vivo generation of secondary NPs [34]. Mesozooplankton, such as copepods and cladocerans, ingest MPs that mimic algal prey in size and optical properties. Biofilm-coated particles emit dimethyl sulfide-like cues, chemically mimicking natural food sources and enhancing trophic uptake.

Suspension and filter feeders – including bivalves, tunicates, and polychaetes – capture particles non-selectively during filtration. In *Mytilus edulis*, MPs intercepted on mucus-covered gill filaments are entrained by ciliary currents and directed to labial palps for ingestion. Particle retention efficiency depends on mucus viscosity, zeta potential, and hydrodynamic shear [35]. Large MPs may be expelled as pseudofeces, whereas NPs penetrate epithelia via paracellular diffusion or vesicular transcytosis.

Planktivorous fish and larval amphibians ingest MPs < 2 mm, particularly under turbid or low-light conditions that reduce visual prey discrimination [35]. Benthic and deposit-feeding invertebrates, such as oligochaetes and amphipods, ingest particles incidentally while foraging, retaining them for extended gut residence times, which enhances leachate absorption and gut–epithelial interaction.

Trophic transfer serves as a secondary exposure pathway. Predators consuming pre-exposed prey can acquire NPs either directly from gut contents or via

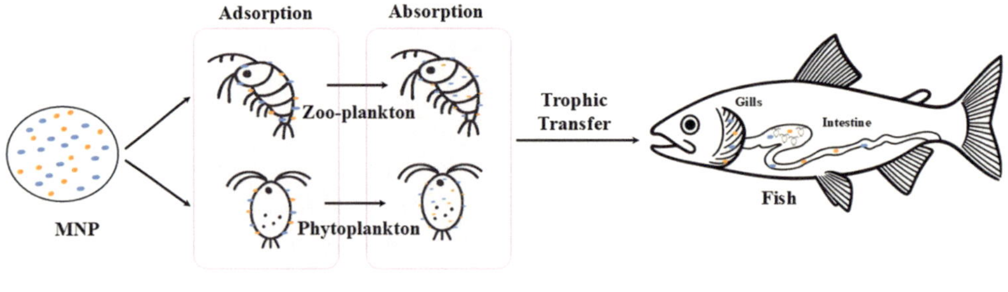

Figure 6.
Trophic transfer of MNPs across aquatic food webs. MNPs adsorb or are absorbed onto/by phytoplankton or zooplankton, which serve as the primary carriers in aquatic food chains. Through predator–prey interactions, these particles are transferred to higher trophic levels, including fish, where they localize in critical organs such as gills and intestines.

liberation from tissues during digestion. Larval zebrafish fed MP-laden zooplankton exhibit hepatic and neural accumulation, demonstrating cross-trophic particle transfer [36].

Surface-based entry routes complement ingestion. NPs adhere to gills, skin, mucosa, and egg chorions, leveraging colloidal stability and small size. In fish gills, NPs bind to mucopolysaccharide layers and penetrate via clathrin- and caveolin-mediated endocytosis, entering lamellar vasculature. In zebrafish embryos, NPs adhere to the chorion and translocate into the yolk sac and developing tissues prior to the onset of active feeding [37].

Benthic invertebrates with permeable cuticles – such as polychaetes and cnidarians – absorb NPs through direct epithelial contact in porewater. *Hydra vulgaris* demonstrated tentacle-bound NP uptake via epithelial phagocytosis, localizing to intracellular vesicles, a process facilitated by negative surface charge and biofilm-mediated membrane fusion [38].

Environmental context modulates uptake efficiency. Turbid, nutrient-rich waters enhance particle adherence via extracellular polymeric substances (EPS), whereas oligotrophic or high-salinity environments reduce mucosal hydration and suppress surface attachment. Field-collected shrimp and crustaceans often display substantial gill-associated plastic burdens, even in gut-empty individuals, confirming non-ingestive uptake pathways [39].

Host physiology influences susceptibility. Larval stages exhibit higher plastic burdens due to increased metabolic demand, permeable epithelia, and immature immune defenses, while inflamed or stressed tissues promote NP internalization via enhanced endocytosis and compromised junctions.

3.2 The role of biofilms and organic coatings

Upon release, synthetic particles undergo rapid physicochemical metamorphosis, acquiring an eco-corona – a dynamic coating of organic matter and microbial biofilms – which redefines their biological identity. This multilayered interface, composed of EPS, DOM, microbial assemblages, and colloidal ions, transforms inert plastics into biologically interactive vectors capable of trophic assimilation and intracellular trafficking.

Molecular architecture of the eco-corona is heterogeneous, dictated by polymer type, environmental chemistry, and microbial succession (**Figure 4**). EPS from Pseudoalteromonas, Shewanella, and Vibrio species is rich in uronic acids, sulfated polysaccharides, and phosphorylated glycoproteins, which modulate zeta potential, hydrophobicity, and aggregation kinetics [40].

Eco-coronas enhance biological recognition and adhesion. AFM and XPS studies reveal that corona-coated NPs adhere more strongly to gill and intestinal epithelia than virgin particles. In larval zebrafish, corona formation accelerates PS NP internalization by increasing clathrin-mediated endocytic activity [37]. Biofilm-associated chemical cues, including dimethyl sulfide analogs, mimic algal signals, misleading copepods and filter feeders into ingestion.

Biofilms also modulate particle retention and translocation. In *Mytilus edulis*, biofilm-coated MPs adhere more efficiently to gill mucus, resist clearance, and migrate into digestive diverticula, with 72-hour biofilms yielding 2.8 × higher translocation than uncolonized particles [41]. In shrimp (*Litopenaeus vannamei*), biofilm-conditioned MPs showed 60% higher hepatopancreatic retention, elevated oxidative stress, and lysosomal destabilization [39].

Eco-coronas also act as chemical concentrators. EPS networks chelate Cd^{2+}, Pb^{2+}, antibiotics, and hydrophobic organics via π–π and ionic interactions. In marine microcosms, Alteromonas-colonized PS MPs adsorbed > 3 × more benzo[a]pyrene than abiotic controls, and corona presence extended desorption half-life in simulated gut fluid from 6 to 19 h [42].

The synergy between plastics, biofilms, and sorbed pollutants amplifies toxicological outcomes. *M. galloprovincialis* exposed to biofilm-coated PVC exhibited vitellogenin synthesis and estrogen receptor activation, primarily due to phthalate leachates and microbial metabolites [43].

Moreover, eco-coronas stabilize particles in prey, facilitating trophic retention and vertical transfer. Rotifers and amphipods carrying biofilm-coated particles maintain gut loads long enough for predators to ingest contaminated prey. Plastisphere biofilms further reshape microbial communities and biogeochemical cycles, enriching denitrifiers and sulfate reducers, which enhance N_2O and H_2S fluxes and impair primary producer biofilms [44].

Thus, the eco-corona is not incidental – it is a primary determinant of plastic bioavailability, toxicity, and trophic mobility, transforming particles into ecologically active microhabitats and contaminant vectors.

3.3 Size-based differences in uptake: MPs vs. NPs

The dimensionality of synthetic particles dictates their biological interaction regime, with MPs confined largely to extracellular compartments and NPs penetrating intracellular and organ-specific niches. NPs leverage curvature-driven membrane affinity and electrostatic interactions to cross epithelial barriers. They utilize clathrin- and caveolin-mediated endocytosis, macropinocytosis, and occasionally direct lipid bilayer penetration. In *Perna viridis* hemocytes, 50-nm PS NPs exhibited ~ 10 × greater granulocyte uptake than 200-nm NPs, demonstrating size- and charge-dependent phagocytic specificity [44].

Size-dependent retention and tissue disruption are evident in vertebrates. In *Oreochromis niloticus*, ~ 86 nm NPs induced more pronounced intestinal damage and longer retention than ~ 185 nm NPs, indicating a retention-size trade-off [45]. Zebrafish embryos exposed to 20 nm PS NPs exhibited transchorionic penetration and couptake of phenanthrene, confirming NP-mediated contaminant transport [46]. MPs, by contrast, are generally restricted to mucosal surfaces – gill epithelia and gut lumen – without systemic translocation. Field observations show size-selective ingestion: Zooplankton rarely internalizes MPs > 300 μm, while benthic feeders accumulate 300–3000 μm particles in digestive tissues [47, 48].

NPs demonstrate organotropism, preferentially accumulating in the liver, gonads, and brain, bypassing barriers inaccessible to MPs. After endocytic uptake, some NPs escape lysosomal degradation, entering cytosolic and even nuclear compartments, conferring higher systemic and reproductive risk potential [49].

4. Cellular and molecular toxicodynamics

4.1 Intracellular trafficking and endocytic pathways

NPs, particularly < 200 nm, exploit multiple endocytic routes that define their cellular uptake, subcellular distribution, and toxicodynamic outcomes. Among these, clathrin-mediated endocytosis (CME) is a dominant entry route. NPs engage membrane-bound receptors, become encapsulated in clathrin-coated vesicles, and traffic to early endosomes, where sorting dictates progression toward late endosomes and lysosomes (**Figure 7**). In bivalve hemocytes, 50–100 nm PS NPs colocalized with LAMP1-positive lysosomes, with granulocytes exhibiting higher uptake than agranulocytes, confirming size- and cell-type-dependent CME [50].

Particles < 100 nm may also utilize caveolae-mediated endocytosis, a pathway associated with cholesterol-rich microdomains and lysosome bypass. Caveolar vesicles preferentially traffic toward non-acidic compartments, including the Golgi and ER, enabling prolonged intracellular retention and retrograde transport. Comparative studies revealed that 50 nm NPs show lower lysosomal colocalization

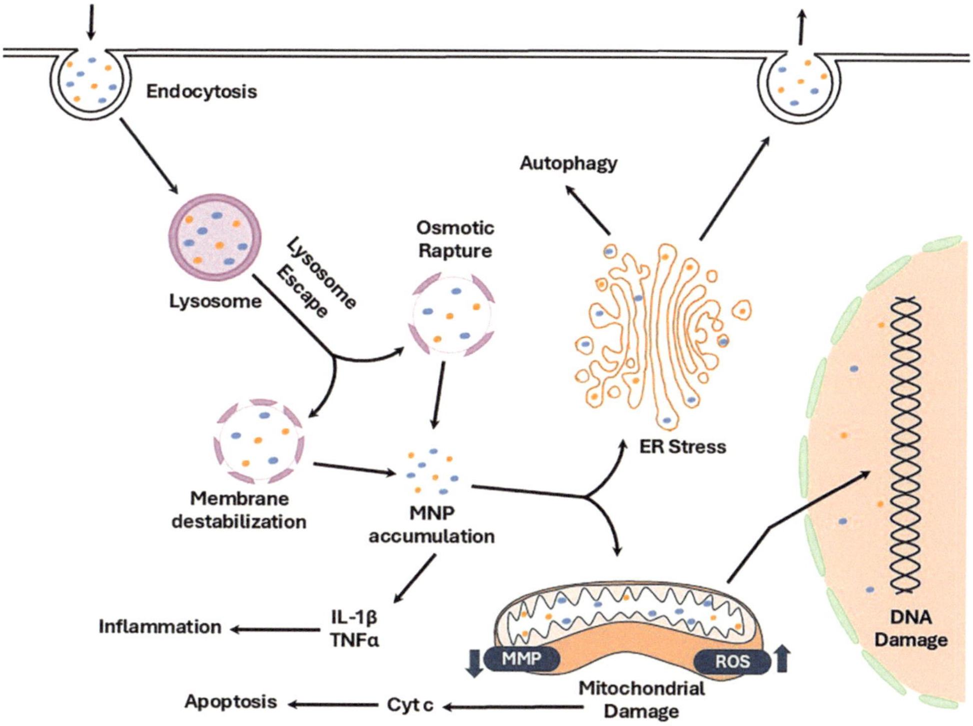

Figure 7.
Cellular pathways of MNPs-induced toxicity. Following endocytosis, MNPs accumulate within lysosomes, where membrane destabilization, osmotic rupture, or lysosomal escape redirects particles into the cytoplasm. Intracellular accumulation triggers oxidative stress and mitochondrial dysfunction, characterized by ROS generation, cytochrome c (Cyt c) release, and apoptosis. Concurrently, MNPs induce endoplasmic reticulum (ER) stress, autophagic disruption, and DNA damage, activating pro-inflammatory mediators such as IL-1β and TNFα.

and greater ER association than 200 nm NPs, consistent with caveolar engagement as a lysosome-evading mechanism [51].

Macropinocytosis, an actin-driven, non-specific engulfment pathway, dominates in phagocytic and immune cells. In *Perna viridis* hemocytes, 100 nm PS NPs localized to multivesicular and multilamellar structures characteristic of autophagic intermediates, indicating a vesicular bottleneck and lysosomal saturation as a driver of sublethal stress [52].

Direct membrane penetration occurs for ultrasmall (<50 nm), hydrophobic, or cationic NPs, which can disrupt phospholipid bilayers and translocate into the cytosol. Carbon-based NPs, visualized via real-time multispectral tracking, demonstrated loss of acidic vesicle colocalization, suggesting endosomal escape and cytosolic dispersal to perinuclear regions [53]. Surface charge strongly modulates uptake and fate. Cationic NPs display enhanced membrane affinity and CME-dominant trafficking, with a fraction escaping vesicular confinement and associating with mitochondria. Serum coronas partially mask charge but preserve receptor-mediated uptake [54].

Protein coronas fundamentally alter recognition and routing. Adsorbed proteins (e.g., albumin, LDL, apolipoproteins) can mimic endogenous ligands, trigger specific endocytosis, and influence lysosomal routing. In endothelial cells, coronated NPs exhibited subdiffusive vesicular motion with low lysosomal bypass probability [55]. Cell type determines intracellular fate. Epithelial cells in gills and gut typically channel NPs to lysosomes, limiting systemic spread. Neurons, embryonic cells, and immune populations, with high endocytic activity or incomplete vesicle maturation, allow lysosomal evasion. Synaptic vesicle recycling has been proposed as a mechanism for neural NP entry, implicating potential neurotoxicity in aquatic species.

Lysosomal accumulation triggers volumetric swelling, pH collapse, and membrane destabilization, releasing cathepsins and hydrolases into the cytosol, which activates inflammasomes and apoptotic signaling. Chronic exposure disrupts autophagic flux, leading to vesicle congestion, organelle stress, and impaired cellular clearance capacity.

4.2 Oxidative stress and organelle damage

Intracellular MNPs disrupt cellular redox homeostasis through surface-driven catalysis, organelle interactions, and chemical leaching, triggering ROS generation and organelle-specific injury. ROS production arises from MNP surface chemistry – notably weathered or aromatic-rich NPs with redox cycling capacity – and from secondary cellular responses. Phthalates and bisphenols leached from MNPs exacerbate ROS by interfering with mitochondrial electron transport and NADPH oxidase activity [56].

Lysosomal destabilization liberates redox-active iron, fueling Fenton reactions and amplifying oxidative bursts. Endosomal escape enables direct mitochondrial contact, disrupting complexes I–III of the ETC, sustaining ROS overproduction and bioenergetic collapse [57]. Lipid peroxidation is an early hallmark, generating reactive aldehydes (4-HNE, MDA) that compromise membrane fluidity and initiate apoptotic cascades [58]. Protein carbonylation and thiol oxidation follow, impairing structural proteins, ion transporters, and metabolic enzymes such as ATP synthase and aconitase, culminating in cellular energy deficits [59]. DNA damage manifests as 8-oxo-dG accumulation, single-/double-strand breaks, and γ-H2AX foci formation, activating p53-dependent repair and apoptotic pathways. Antioxidant defenses are rapidly depleted: GSH levels drop, and superoxide dismutase (SOD) and CAT activities decline, reflecting both overconsumption and transcriptional suppression [58].

Mitochondria are central toxicity targets. MNPs induce $\Delta\Psi$m loss, Cyt c release, and intrinsic apoptosis activation [58, 60]. Oxidative phosphorylation collapses, with ETC complex downregulation and ATP depletion, prompting a glycolytic shift (Warburg-like response) [58]. Morphological hallmarks – swelling, cristae loss, and fragmentation – intensify under NP–metal coexposures, accelerating mitophagy [57]. Calcium dysregulation exacerbates mitochondrial injury. Aberrant ER–mitochondria Ca^{2+} transfer triggers mPTP opening and permeability transition, precipitating organelle collapse [59]. ER stress is a concurrent axis of injury, evidenced by UPR activation (PERK, eIF2α, CHOP upregulation) in mollusks and fish transcriptomes [61]. Lysosome–mitochondria crosstalk links cathepsin release to intrinsic apoptosis, and autophagic flux becomes incomplete, leading to persistent organelle damage and redox imbalance [62, 63].

4.3 Genotoxic and epigenetic impacts

MNPs exert genotoxicity through physical nuclear interactions, oxidative DNA damage, and epigenetic remodeling, with long-term developmental and transgenerational implications. Particles < 100 nm can penetrate nuclei directly or via endosomal escape, especially in highly mitotic or gonadal cells. In *Corbicula fluminea*, nuclear-localized NPs caused DNA fragmentation and mitochondrial apoptosis, effects absent in larger MPs [64]. Mitochondria-derived ROS induce 8-oxo-dG lesions and double-strand breaks, triggering γ-H2AX formation and p53-dependent apoptosis.

Transcriptomic reprogramming follows initial DNA stress. *Daphnia pulex* exposed to PS-NPs showed repression of chitinase and trehalose metabolism genes and induction of reproductive regulators (doublesex, hormone-related genes). Zebrafish embryos exhibited p53, Bax/Bcl-2, and caspase activation, coupling DNA damage sensing with mitochondrial apoptosis and developmental perturbation [65]. Epigenetic remodeling represents a critical but often silent layer of MNP toxicity. Global hypomethylation with promoter-specific hypermethylation disrupts chromatin architecture and silences key developmental and tumor suppressor genes. Histone marks shift toward H3K27me3 (repressive) and H3K9ac (activating), altering chromatin accessibility. Histone-modifying enzymes (EZH2, HDAC1) respond dynamically to carbonaceous and PS-NPs [66].

miRNA dysregulation adds a post-transcriptional layer to NP-induced toxicity. miR-21, miR-125, and miR-34 are differentially expressed, linking redox stress to NF-κB activation and p53-mediated apoptosis, amplifying cellular injury and organ dysfunction. Germline and developmental toxicity are ecologically pivotal. Zebrafish and mollusk exposures during early cleavage and organogenesis induce chromosomal bridges, abnormal morphogenesis, and reduced hatching [67]. Gonadal NP accumulation leads to micronuclei formation, meiotic errors, and epigenetic shifts, raising concern for transgenerational inheritance of mutational burdens. Particle physico-chemistry dictates genomic risk. Cationic and < 50 nm NPs exhibit enhanced nuclear affinity and DNA interaction, while protein coronas mimicking albumin or LDL can facilitate nuclear pore entry and transcriptional interference [68].

5. Physiological disruption in aquatic species

5.1 Neurological and behavioral impairments

NPs exert profound neurophysiological and behavioral effects in aquatic species due to their unique ability to traverse biological barriers and directly localize within neural tissues (**Figure 8**). Unlike MPs, which predominantly act peripherally or elicit secondary systemic effects, NPs can penetrate the blood–brain barrier via transcellular transport or olfactory neuronal pathways, allowing them to access privileged compartments of the central nervous system. In zebrafish, fluorescent polystyrene NPs of approximately 70 nm were observed to distribute across the telencephalon, cerebellum, and optic tectum, with ultrastructural evidence of neuronal somatic accumulation and glial activation, indicating early signs of neuroinflammation [69]. Similar patterns of neural deposition have been reported in mollusks and nematodes, supporting a conserved vulnerability across aquatic taxa [70].

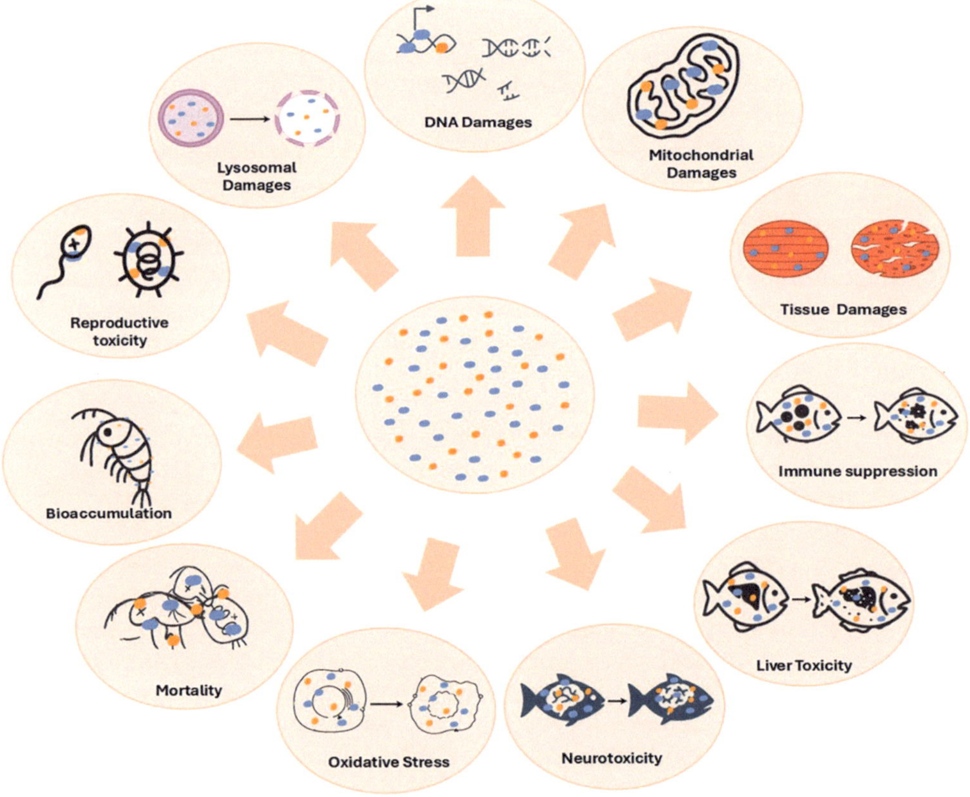

Figure 8.
Multiscale toxicological impacts of MNPs on aquatic organisms. MNP exposure induces a spectrum of adverse outcomes, including lysosomal destabilization, DNA damage, mitochondrial dysfunction, and tissue-level injury. These cellular disruptions manifest as oxidative stress, immune suppression, liver toxicity, and neurotoxicity, ultimately impairing reproduction, promoting bioaccumulation, and increasing mortality.

At the cellular level, NP exposure induces synaptic degeneration, dendritic spine retraction, and the downregulation of presynaptic markers such as synaptophysin and PSD-95, which collectively impair neurotransmission and diminish neuroplasticity. Neurodevelopment is especially sensitive to NP insult. Zebrafish embryos exposed to polystyrene NPs exhibit disrupted axonal patterning, reduced expression of neurodevelopmental regulators, including NeuroD1 and GFAP, and impaired differentiation of neural progenitor zones, which result in long-lasting locomotor deficits and aberrant sensorimotor integration post-hatching.

Neurological impairment is closely linked to molecular disturbances in neurotransmitter homeostasis. NP exposure consistently reduces dopamine (DA) and serotonin (5-HT) levels, correlating with movement dysregulation, aggression, and anxiety-like phenotypes. Cholinergic signaling is also disrupted, as evidenced by suppressed acetylcholinesterase (AChE) activity, which compromises motor coordination and short-term learning. These biochemical changes are often exacerbated by oxidative stress, which destabilizes neuronal membranes, disrupts calcium homeostasis, and induces excitotoxic damage.

These subcellular and molecular disruptions manifest as overt behavioral phenotypes, including hyperactivity, hypoactivity, heightened thigmotaxis, reduced exploratory behavior, and impaired habituation and associative learning [71]. Such behavioral deviations extend to ecologically critical functions. Larval and juvenile fish, rotifers, and copepods with NP accumulation display prolonged escape response latencies and impaired predator recognition, while olfactory and visual processing deficits compromise both prey detection and predator evasion. This neuroethological disruption diminishes foraging efficiency, shoaling dynamics, and mating performance, leading to a cumulative decline in ecological fitness under chronic NP exposure.

5.2 Immune system modulation

The immune system of aquatic organisms is a major target of MNPs, which engage innate and adaptive components through size-, surface-, and charge-dependent interactions (**Figure 8**). NPs act as immunologically active particles capable of mimicking pathogen-associated molecular patterns (PAMPs) and triggering dysregulated immune responses. In zebrafish larvae, coexposure to PS-NPs (~100 nm) and MPs (5 μm) activates the TLR4/MyD88 signaling pathway, producing robust upregulation of pro-inflammatory cytokines such as IL-1β and TNFα, while simultaneously suppressing interferon regulatory genes that are essential for antiviral immunity [72]. Renal accumulation of NPs in teleosts, where hematopoiesis occurs, indicates organ-specific immunosuppression with potential systemic consequences.

Surface chemistry and biofilm formation play pivotal roles in immune modulation. Oxidized or biofilm-coated NPs present molecular patterns that are readily recognized by toll-like receptors (TLRs) and scavenger receptors on hemocytes and macrophages. In *Ciona robusta*, 50-nm PS-NPs altered the transcription of TLR and stress-related genes in the pharyngeal immune interface, while simultaneously suppressing LPS-induced immune memory, indicating impaired innate priming and potential loss of trained immunity in sessile invertebrates [73].

Cellular and enzymatic immune defenses are similarly perturbed. In mussels, NP exposure disrupts hemocyte homeostasis by altering the hyalinocyte-to-granulocyte ratio and induces a transient elevation in SOD activity during the first 24 hours of exposure, reflecting acute oxidative activation followed by functional exhaustion [74]. In teleosts, chronic MP exposure downregulates key antimicrobial peptides, including hepcidin, defensin, and lectins, as well as enzymatic effectors such as lysozyme and peroxidase, which weakens mucosal barrier immunity and impairs bacterial clearance. Histopathological evidence from zebrafish and other species demonstrates splenic lymphoid depletion, vacuolization in the head kidney, and erythrocyte membrane damage under chronic exposure. In gut-associated lymphoid tissue, NP exposure leads to suppression of TLR4, IL-8, and IL-1β expression, accompanied by microbiome dysbiosis, which disrupts mucosal barrier integrity and systemic immune homeostasis [75]. These factors may contribute to lymphoma development, as proposed in our recent review [8].

Mechanistically, immune disruption is driven by lysosomal stress and ROS production [76]. Internalized NPs localize within acidic lysosomal compartments, where they destabilize membranes and release cathepsins, activating inflammasome pathways, likely including NLRP3, and triggering pyroptotic-like cell death when combined with viral mimics [77, 78]. Macrophage models demonstrate that NPs activate the cGAS–STING cytosolic DNA-sensing pathway, resulting in NF-κB nuclear translocation and type-I interferon gene induction, with activation magnitude influenced by particle surface charge and protein corona composition [78]. Chronic overactivation leads to sterile inflammation, tissue damage, and autoimmunity-like pathologies. Comparative studies reveal a size-dependent dichotomy: Larger MPs can transiently enhance immune vigilance, whereas NPs often suppress immune gene expression and antioxidant defenses, increasing pathogen susceptibility, as demonstrated in *Bombyx mori* exposed to *Serratia marcescens* [79]. Biofilm-coated MPs further exacerbate immune dysregulation by acting as vectors for opportunistic pathogens such as Vibrio spp., which adhere to mucosal surfaces, breach epithelial barriers, and potentiate inflammation in gill and gut tissues, effectively functioning as Trojan horse complexes.

5.3 Reproductive and developmental toxicity

Reproductive impairment caused by MNPs is mechanistically multifaceted, encompassing endocrine disruption, gonadal architecture damage, germline compromise, and embryonic toxicity (**Figure 9**). These particles perturb the hypothalamic–pituitary–gonadal axis by altering sex steroid homeostasis and feedback regulation. In *Danio rerio*, exposure to PS-NPs led to reduced plasma estradiol (E2) and elevated testosterone, consistent with aromatase inhibition and feedback suppression of steroidogenesis [80]. Correspondingly, transcription of ovarian genes cyp19a1a, ERα, and LHR was downregulated, impairing folliculogenesis, luteinization, and oocyte maturation. Male teleosts exhibited analogous dysfunction, with disrupted Leydig and Sertoli cell function, seminiferous tubule degeneration, and sperm maturation defects [81]. These hormonal and cellular effects are exacerbated by the leaching of phthalates, bisphenols, and flame retardants from particle surfaces, which act as endocrine disruptors by activating estrogenic pathways or antagonizing androgenic signaling [82].

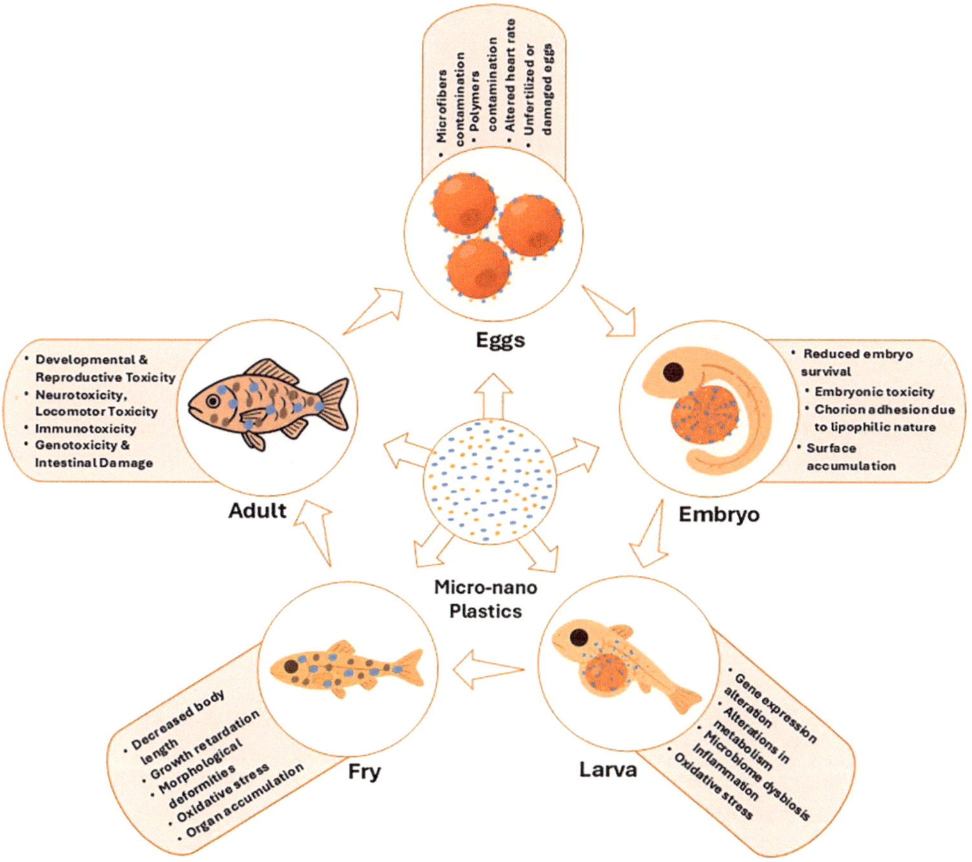

Figure 9.
Life-stage-specific effects of MNPs in fish. MNP exposure exerts stage-dependent toxicity across the developmental continuum. In eggs, particles adhere to the chorion, alter fertilization success, and reduce hatching rates. Embryos display impaired survival, embryotoxicity, and surface accumulation due to lipophilic adhesion. Larvae show altered gene expression, disrupted metabolism, microbiome dysbiosis, inflammation, and oxidative stress. Fry exhibit decreased growth, morphological deformities, and organ-level accumulation. Adults experience developmental and reproductive toxicity, neurotoxicity, immunotoxicity, genotoxicity, and intestinal injury.

Histopathological evaluations across multiple aquatic taxa reveal extensive gonadal degradation. In females, MPs induce follicular atresia, delayed vitellogenesis, granulosa cell apoptosis, and eventual fibrotic remodeling of the ovarian stroma [83]. In males, particle accumulation in the testicular interstitium compromises the blood–testis barrier, reduces spermatocyte proliferation, and results in malformed sperm with mitochondrial cristae loss and flagellar defects, contributing to impaired motility [84]. At the cellular level, oxidative stress drives much of this reproductive toxicity by destabilizing lipid membranes, damaging organelles, and triggering apoptosis in Leydig and Sertoli cells, with INSL3 downregulation serving as a biomarker of hormonal collapse [81].

Embryonic and early developmental toxicity is a downstream consequence of parental exposure. Female zebrafish exposed to PS NPs exhibit reduced fecundity, fewer mature oocytes, and offspring (F1) with increased heart rate, delayed hatching, and dysregulated sex hormone levels, confirming maternal transfer of particle-induced stress [80]. Key developmental regulators, such as sox9a and foxl2a, are disrupted, producing defects in sex differentiation, neurodevelopment, and organogenesis. Embryo-larval assays consistently document yolk sac edema, craniofacial malformations, and growth retardation, with severity scaling with particle size reduction and surface reactivity [82].

5.4 Transgenerational effects and chronic exposure outcomes

The physiological impacts of MNPs extend beyond directly exposed individuals, propagating through germline transmission and epigenetic programming to produce persistent, multigenerational effects. NPs penetrate deep into reproductive tissues, including the blood–testis barrier, ovarian granulosa layers, and embryonic chorion analogs. Microscopic confirmation of NP bioaccumulation in sperm heads, ooplasm, and vitelline envelopes has been reported in *Danio rerio* and *Daphnia magna*, while *C. elegans* exposed to photoaged PS-NPs demonstrated germline localization associated with reduced brood size and intergenerational fecundity decline [85].

Multigenerational studies reveal that MP and NP exposure can induce epigenetic alterations that persist beyond the exposure period. In fathead minnows, DNA methylation shifts were observed in regulatory elements controlling endocrine and xenobiotic response pathways, persisting into the F1 generation [86]. In *C. elegans*, heritable histone modification patterns, including H3K9me2 and H3K27me3, silenced fertility-critical genes such as PIE-1 and GLP-1 across three generations [85]. The deregulation of stress-sensitive miRNAs, notably miR-34 and miR-21, in germline tissue further mediates intergenerational transmission of apoptotic sensitivity and developmental misregulation.

Phenotypic consequences are persistent and ecologically significant. F1 zebrafish from NP-exposed parents exhibit spinal curvature, swim bladder malformation, reduced fertilization rates, and diminished larval viability, even after maternal particle clearance, indicating stable inheritance of toxic cues [87]. Sex ratio skewing toward males has been observed in Daphnia and zebrafish, reflecting suppression of estrogen signaling across generations. Behavioral impairments in progeny include reduced exploration, impaired predator evasion, and abnormal courtship, which correlate with dopaminergic dysregulation and maternal cotransfer of adsorbed pollutants such as TDCIPP [88].

Chronic and life-cycle exposures culminate in systemic organ compromise. Mitochondrial swelling, cristae loss, and hepatic and gonadal energy collapse underpin reproductive decline, while immune organs exhibit delayed healing, persistent inflammation, and heightened pathogen susceptibility [89]. Even after exposure cessation, oxidative memory encoded within mitochondrial and nuclear compartments persists, impeding physiological recovery. Experimental interventions, such as melatonin supplementation, partially restore mitochondrial respiration and reproductive performance in NP-exposed zebrafish, implicating ROS generation as a central mechanistic trigger in chronic, multigenerational pathophysiology [87].

6. Trophic dynamics and ecosystem feedback

6.1 Evidence of food web contamination

At the base of aquatic food webs, phytoplankton and microalgae represent the first biological interfaces for MNP contamination. NPs exploit their small size, high surface-area-to-volume ratios, and electrostatic interactions to adhere to algal cell walls, embedding within extracellular matrices and forming nanoscale coronas. Transmission electron microscopy of Chlorella and Scenedesmus has revealed PS-NPs physically lodged in the cell wall matrix, where they disrupt ion exchange and nutrient uptake, while localized ROS generation damages photosystem II and interferes with Calvin cycle fluxes. This biochemical interference reduces chlorophyll a synthesis and primary productivity, producing energetic deficits that propagate upward into zooplankton and higher consumers. Importantly, these effects have been documented even at environmentally realistic particle concentrations, highlighting that the basal energy engine of aquatic ecosystems is chronically exposed to sublethal plastic interference long before particles reach higher trophic levels.

Zooplankton constitute the first active kinetic vector for plastic entry into the food web. Their non-selective suspension feeding and chemosensory-driven foraging render them susceptible to particle–prey mimicry. Field-collected copepods and rotifers ingest MPs in the 2–50 μm range, while submicron NPs evade mechanical rejection and penetrate gut epithelia via paracellular diffusion or endocytosis. Laboratory tri-trophic experiments with Brachionus–Mesocyclops–sea bass fry confirm that MPs are ingested at rates of up to 3.7 particles per individual per minute, suppressing algal clearance rates and fecundity [90]. Field evidence mirrors these laboratory trajectories: in Mexican tropical lakes and Phuket coastal waters, copepods, fish larvae, and bivalve veligers frequently harbor MNPs in their guts, with particle loads fluctuating seasonally and increasing near anthropogenic discharge zones [91, 92].

Parallel contamination occurs in benthic compartments, where MPs settle into sediments and interact with sessile or deposit-feeding invertebrates. *Mytilus galloprovincialis* filters MPs via ciliary and mucus-bound pathways, trapping particles in gill lamellae and digestive diverticula, where hemocyte uptake of NPs initiates ROS production, lysosomal destabilization, and immune suppression [93]. Deposit feeders and polychaetes ingest sediment-associated MPs that are frequently biofilm-coated and adsorbed with heavy metals and PAHs, creating a synergistic exposure regime at the sediment–benthos interface. This dual stressor model – physical particle abrasion coupled with chemical vectoring – renders the benthic zone a potent reservoir of combined plastic and pollutant toxicity.

Planktivorous and benthivorous fish integrate both pelagic and benthic contamination pathways. In Chirostoma spp., MPs are routinely recovered from gastrointestinal tracts at burdens of 4–5 particles per individual, correlating with local plankton ingestion rates and confirming active trophic coupling [91]. Mesocosm studies show that zooplankton-mediated NP transfer leads to organotropism in larval fish: Particles bypass gut egestion, translocate to liver and neural tissues, and induce metabolic and behavioral disruptions [94]. Field surveys in Jiaozhou Bay extend this pattern to carnivorous chaetognaths, which accumulate MPs via copepod prey, providing real-world evidence of stepwise trophic transfer [95].

This contamination cascades to apex consumers such as seabirds and marine mammals. Fulmars, petrels, and albatrosses ingest MPs indirectly through contaminated prey, resulting in sublethal malnutrition, impaired chick survival, and occasional gastrointestinal blockage [96].

6.2 Bioaccumulation vs. biomagnification

The trophic behavior of MNPs diverges from that of lipophilic Persistent Organic Pollutants. Bioaccumulation is frequently observed at the organismal level when ingestion exceeds egestion, but classical biomagnification across trophic levels remains rare. Field investigations in Chinese coastal waters revealed that the demersal fish *Collichthys lucidus* retained an average of 3.24 ± 1.57 MPs per individual, primarily blue PET fibers, with higher burdens in older, benthivorous individuals. In contrast, sympatric *Larimichthys polyactis* showed minimal retention, illustrating strong species- and behavior-dependence [97]. Freshwater observations from the Three Gorges Reservoir documented 0.68–4.00 MPs per fish, largely < 1 mm fibers, with low bioaccumulation factors because most particles remained confined to gut contents rather than edible tissues [98].

Trophic transfer of MNPs is well-documented, but biomagnification is uncommon because particles are usually egested, remain gut-bound, and do not partition into lipid phases. Coral reef and coastal food chain studies illustrate this phenomenon: a tri-level chain of zooplankton → benthic crustaceans → reef fish exhibited abundant particle presence in basal taxa but trophic magnification factors near 1.0, indicating negligible biomagnification [99]. British Columbia field surveys also showed the highest particle prevalence in small planktivores and filter feeders, with lower relative burdens in top predators, suggesting trophic dilution [100]. Modeled cetacean food webs indicate that while zooplankton and forage fish exhibit measurable particle loads, whales do not biomagnify plastics due to efficient digestive clearance [101].

Mechanistically, the scarcity of biomagnification reflects particle-scale constraints and physiological clearance. MPs above 100 μm remain gut-limited and are typically expelled within 24–96 hours. NPs and submicron MPs, however, can penetrate epithelia, localize in the liver, gonads, or neural tissue, and persist in lysosomal compartments, creating localized bioaccumulation that is often invisible to gut content surveys. Feeding ecology further mediates exposure: Sessile filter and deposit feeders experience chronic particle contact, while mobile predators encounter intermittent pulses. Unlike hydrophobic contaminants, MPs lack lipid affinity, precluding classical chemical magnification. Functional biomagnification may occur indirectly if NPs vector adsorbed PAHs or metals into tissue, enhancing sublethal stress without a numerical increase in particle counts.

6.3 Community shifts and functional declines

Chronic exposure to MNPs propagates beyond individuals to reshape aquatic community structure and ecosystem function. At the organismal level, these particles act as both physical stressors and biochemical disruptors. Zooplankton and larval fish experience caloric dilution as ingested MPs occupy gut volume without nutritional benefit, leading to growth suppression and reduced reproductive output. NPs exacerbate this by penetrating gut epithelia, where they induce mitochondrial

ROS production, lysosomal destabilization, and epithelial apoptosis. These cellular injuries manifest as diminished grazing efficiency, lower fecundity, and heightened predation risk.

In benthic systems, filter feeders and sediment bioturbators, such as mussels and polychaetes, internalize MPs and phagocytose NPs into hemocytes, triggering lysosomal membrane permeabilization, ATP depletion, and immune compromise. Functional consequences include reduced sediment reworking, burrowing activity, and biofiltration, leading to early community simplification characterized by keystone species decline and opportunist proliferation [102].

Primary productivity and nutrient cycling are also disrupted. MPs adhering to phytoplankton surfaces create shading effects, while NPs entering chloroplasts impair photosystem II electron transport and initiate ROS-mediated lipid peroxidation, suppressing carbon fixation. MPs rapidly acquire biofilms that sequester dissolved organic carbon toward heterotrophic bacteria, reducing autotrophic competitiveness and shifting microbial community structure. Mesocosm studies have linked these interactions to decreased primary productivity, altered nitrogen and phosphorus cycling, and early destabilization of aquatic stoichiometry [103, 104]. Energy bottlenecks at the base of the food web cascade upward, reducing prey quality and quantity for herbivores and predators.

Ecosystem-wide, these interactions converge to diminish functional performance. Predatory fish exhibit energy deficits as MP-laden prey deliver lower caloric value, while NP translocation to the liver, gonads, and neural tissue induces chronic metabolic and neuroendocrine stress. In benthic communities, impaired bioturbation slows sediment oxygenation and denitrification, and reduced bivalve filtration diminishes water clarity and organic matter turnover [105]. Freshwater biofilms exposed to PS-NPs demonstrate downregulated pigment gene expression, reduced extracellular enzyme activity, and fragmented microbial interaction networks, reflecting early loss of ecosystem multifunctionality. These organismal and community-level disruptions aggregate into biodiversity loss, trophic flattening, and ecosystem homogenization, leaving aquatic systems energy-inefficient and highly susceptible to secondary stressors such as eutrophication, hypoxia, and warming.

6.4 Cross-ecosystem transport (freshwater to marine)

The migration of MNPs from freshwater to marine environments constitutes a multistage process driven by hydrology, particle physicochemistry, sediment dynamics, and ecological vectors. Riverine inputs arise from urban stormwater, wastewater, agricultural runoff, and atmospheric deposition. Under low-flow conditions, plastics largely reside in sediments, floodplains, or lentic deposits, while high-flow events mobilize these reservoirs, transporting particle pulses toward estuaries [106]. Low-density polymers such as PE and PP remain buoyant, entrained in surface waters, whereas denser materials such as PET or biofouled MPs cycle between deposition and resuspension, creating pulsed export events.

Estuarine transition zones act as both particle reactors and retention basins. Salinity gradients, ionic bridging by Ca^{2+} and Mg^{2+}, and interactions with clays and EPS promote heteroaggregation and flocculation, driving some particles into benthic muds while others persist as colloidal plumes [107]. Biofilm formation accelerates in brackish conditions, altering particle density and surface reactivity, while tidal pumping and anthropogenic dredging episodically remobilize settled plastics for export to coastal seas [108].

During this cross-ecosystem transport, MNPs undergo aging and bioactivation that enhance ecological risk. UV-induced photooxidation generates carbonyl, hydroxyl, and carboxyl groups, increasing hydrophilicity and sorption potential. Mechanical abrasion creates angular fragments with high surface energy, and biofilm colonization forms eco-coronas that concentrate metals, PAHs, and POPs [109]. These transformations increase the probability of epithelial adhesion, mucosal uptake, and trophic transfer.

In the coastal ocean, dispersal is dictated by particle density, biofouling status, and regional hydrodynamics. Buoyant MNPs persist in river plumes and surface gyres for weeks to months before either offshore advection or density inversion via biofouling and mineral bridging. Sinking fractions accumulate in deltaic clinoforms and continental shelf sediments, forming benthic reservoirs that are episodically mobilized by storms or hypoxia. Flood-driven sediment cores from the Yangtze Estuary show sharp MP deposition spikes, linking catchment hydrology directly to coastal loading [110]. Global watershed models indicate that, although only 2–10% of terrestrial plastic inputs ultimately reach the sea, this fraction is sufficient to seed marine food webs continuously, creating persistent benthic and pelagic contamination reservoirs [111, 112].

Ecologically, this freshwater-to-marine flux embeds plastics into cross-ecosystem energy and nutrient networks. Migratory and estuarine-dependent species transport ingested particles and their sorbed contaminants across ecological boundaries, linking river basins to open ocean systems. Estuarine and coastal sediment deposition degrades benthic habitat quality, while persistent NP surface plumes interact with primary producers and microbial consortia, perturbing carbon fixation, nitrogen cycling, and trophic energy transfer. Over time, rivers function not merely as conduits but as dynamic reactors and amplifiers of plastic risk, bridging terrestrial sources with marine sinks and embedding synthetic particles into global biogeochemical cycles.

7. Hidden threats, emerging tools, and strategic roadmaps

7.1 The silent burden of hidden poison

MNPs impose a chronic, cryptic toxicological load on aquatic organisms that rarely manifests as immediate lethality but instead unfolds as a cascade of subcellular injuries with long-term ecological repercussions. MNPs cause oxidative stress and neurobehavioral changes in aquatic organisms, undermining reproduction and fitness, and eventually leading to population declines and ecosystem simplification [113–115]. These sublethal effects accumulate over generations, silently eroding community resilience and ecosystem structure.

The cryptic nature of this threat is compounded by the severe limitations of current detection methods. Standard environmental monitoring tools, such as μ-FTIR and Raman spectroscopy, fail to resolve particles below 10 μm, and even advanced methods are rarely deployed in field surveys [116, 117]. Rapid eco-corona formation camouflages NPs as organic particulates, contributing to false negatives in monitoring and facilitating trophic uptake [118]. In natural ecosystems, the toxicological footprint of MNPs blends with background stressors such as turbidity, hypoxia, and nutrient enrichment, masking their role in long-term ecosystem degradation. Laboratory studies often highlight acute endpoints, but in the field, cumulative sublethal injuries silently erode reproductive success, behavioral integrity, and population resilience. MNPs thus function as a quintessential “hidden poison”: mechanistically active at subcellular scales, ecologically erosive across generations, and effectively invisible to the regulatory and monitoring frameworks that guide aquatic conservation, as the term was recently coined by our laboratory [8].

7.2 Integrative approaches and policy challenges for MNP risk management

Addressing the ecological risks posed by MNPs requires a shift from isolated case studies to predictive, integrated frameworks that connect molecular mechanisms, environmental fate, and ecosystem-level outcomes. Omics technologies now reveal that MNP exposure triggers wide-ranging transcriptional, proteomic, and metabolomic changes in aquatic organisms – such as upregulated oxidative stress genes [119], altered synaptic proteins, and disrupted energy metabolism [120, 121] – linking subcellular injury to reduced reproduction and population decline. Material science tools such as XPS, AFM, FTIR, and TEM have mapped the environmental evolution of MNPs, documenting how UV exposure, mechanical abrasion, and biofilm colonization enhance pollutant sorption and leachate toxicity [122, 123].

Predictive modeling, using quantitative structure–activity relationships and machine learning, now allows hazard forecasting based on particle traits and environmental context [124, 125]. Yet, despite these advances, most regulatory frameworks remain outdated, targeting only macroplastics, lacking standardized MNP definitions, and failing to integrate chronic, sublethal, or generational effects [126, 127]. A coordinated approach is needed: harmonizing definitions, developing particle-specific risk indices, and integrating multi-omics data, mesocosm trials, and in situ monitoring [121, 128].

Ultimately, global governance frameworks – such as the UN Plastics Treaty and the CBD – must adapt to integrate MNP risk, using AI-driven modeling and predictive tools for proactive regulation [129]. Only through the convergence of mechanistic science, advanced monitoring, and harmonized policy can we mitigate the silent, system-wide threat of MNPs to aquatic ecosystems.

Conflict of Interest

The authors declare no conflict of interest.

Use of AI statement

AI assistance, specifically the ChatGPT-4.0 model, was used during the final stages of editing to polish the language. The initial draft was written manually by the author. The AI-generated refinements were carefully supervised and edited by the author, with every line thoroughly reviewed to ensure that the original meaning was preserved and that no unintended changes, omissions, or added emphases were introduced.

Author details

Naveen Kumar[1*], Jayendra Kumar Singh[3], Mrinmoy Mandal[2], Nisha Yadav[2] and Divya Tyagi[2]

1 Assistant Professor, Department of Zoology, School of Basic and Applied Sciences, Sanskaram University, Jhajjar, India

2 Department of Statistics, Institute of Science, Banaras Hindu University (BHU), Varanasi, India

3 Department of Chemistry, School of Basic and Applied Sciences, Sanskaram University, Jhajjar, India

*Address all correspondence to: Naveentak72@gmail.com

References

[1] Pilapitiya PNT, Ratnayake AS. The world of plastic waste: A review. Cleaner Materials. 2024;**11**:100220

[2] Dissanayake PD, Kim S, Sarkar B, Oleszczuk P, Sang MK, Haque MN, et al. Effects of microplastics on the terrestrial environment: A critical review. Environmental Research. 2022;**209**:112734

[3] Chauhan S, Basri S. In: Singh SP, Agarwal AK, Gupta T, Maliyekkal SM. Microplastics in terrestrial soils: Occurrence, analysis, and remediation. New Trends in Emerging Environmental Contaminants. Springer; 2021. p. 67–80

[4] Corti A, La Nasa J, Biale G, Ceccarini A, Manariti A, Petri F, et al. Microplastic pollution in the sediments of interconnected lakebed, seabed, and seashore aquatic environments: Polymer-specific total mass through the multianalytical "PISA" procedure. Analytical and Bioanalytical Chemistry 2023;**415**:2921–2936

[5] Mondellini S, Schott M, Löder MG, Agarwal S, Greiner A, Laforsch C. Beyond microplastics: Water soluble synthetic polymers exert sublethal adverse effects in the freshwater cladoceran Daphnia magna. Science of the Total Environment. 2022;**847**:157608

[6] Mityaev MV, Gerasimova MV, Druzhkova EI. Olymer particles in solid atmospheric precipitation in the northwestern Kola Peninsula in 2020–2021. Russia: Doklady Earth Sciences, Springer. 2022;**505**(2). p. 586–590

[7] Hasan M, Bin MF, Ahmed S, Maksud A, Jubayer A, Rahman A. Microplastics in aquatic ecosystems: Sources, impacts, and challenges for biodiversity, food security, and human health-A meta analysis. Integrative Biomedical Research. 2024;**8**(11):1–12

[8] Kumar N, Pachar AK, Singh JK, Mandal M, Lamba M, Yadav S, et al. Investigating the hypothesis role of the hidden poison microplastics in lymphoma development. Discover Environment 2025;**3**:73

[9] Alqahtani S, Alqahtani S, Saquib Q, Mohiddin F. Toxicological impact of microplastics and nanoplastics on humans: Understanding the mechanistic aspect of the interaction. Frontiers in Toxicology. 2023;**5**:1193386

[10] Vohl S, Kristl M, Stergar J. Harnessing magnetic nanoparticles for the effective removal of micro-and nanoplastics: A critical review. Nanomaterials. 2024;**14**(14):1179

[11] Song J, Wang C, Li G. Defining primary and secondary microplastics: A connotation analysis. ACS ES&T Water. 2024;**4**(6):2330–2332

[12] Wolf C, Wenzel M, Fischer B, Bertling R, Jelen E, Hennecke D, et al. iMulch: An investigation of the influence of polymers on a terrestrial ecosystem using the example of mulch films used in agriculture. Environmental Sciences Europe 2025;**37**:13

[13] Okoffo ED, Thomas KV. Quantitative analysis of nanoplastics in environmental and potable waters by

pyrolysis-gas chromatography–mass spectrometry. Journal of Hazardous Materials. 2024;**464**:133013

[14] An L, Liu Q, Deng Y, Wu W, Gao Y, Ling W. Sources of microplastic in the environment. In: Luo Y, Liu W, editors. Microplastics in Terrestrial Environments: Emerging Contaminants and Major Challenges. Singapore: Springer; 2020. p. 143–159

[15] Dąbrowska A. Marine Microplastics: Chemical, Physical, Biological, and Social Perspectives. Life below Water. Singapore: Springer; 2022. 607–620 p

[16] Napper IE, Thompson RC. Release of synthetic microplastic plastic fibres from domestic washing machines: Effects of fabric type and washing conditions. Marine Pollution Bulletin. 2016;**112**(1-2):39–45

[17] Facciolà A, Visalli G, Pruiti Ciarello M, Di Pietro A. Newly emerging airborne pollutants: Current knowledge of health impact of micro and nanoplastics. International Journal of Environmental Research and Public Health. 2021;**18**(6):2997

[18] El Hadri H, Gigault J, Maxit B, Grassl B, Reynaud S. Nanoplastic from mechanically degraded primary and secondary microplastics for environmental assessments. NanoImpact. 2020;**17**:100206

[19] Dalla Fontana G, Mossotti R, Montarsolo A. Assessment of microplastics release from polyester fabrics: The impact of different washing conditions. Environmental Pollution. 2020;**264**:113960

[20] Zhang H, Fei Y, Wang H, Chen Y, Huang S, Yu B, et al. Interaction of microplastics and organic pollutants: Quantification, environmental fates, and ecological consequences. In: He Defu, Luo Yongming. Microplastics in Terrestrial Environments: Emerging Contaminants and Major Challenges. Cham, Switzerland: Springer; 2020. p. 161–184

[21] Zhao C, Xu T, He M, Shah KJ, You Z, Zhang T, et al. Exploring the toxicity of the aged styrene-butadiene rubber microplastics to petroleum hydrocarbon-degrading bacteria under compound pollution system. Ecotoxicology and Environmental Safety. 2021;**227**:112903

[22] Qin Y, Tu Y, Chen C, Wang F, Yang Y, Hu Y. Biofilms on microplastic surfaces and their effect on pollutant adsorption in the aquatic environment. Journal of Material Cycles and Waste Management. 2024;**26**(6):3303–3323

[23] Chen X, Chen C-E, Cheng S, Sweetman A Bisphenol a sorption on commercial Polyvinyl Chloride microplastics: Effects of UV-Aging and biofilm colonization on plastic behaviour in the environment. Available at SSRN 4763491. 2024

[24] Peng L, Wang Y. Sediment organic carbon dominates the heteroaggregation of suspended sediment and nanoplastics in natural and surfactant-polluted aquatic environments. Journal of Hazardous Materials. 2022;**440**:129802

[25] Liu Y, Huang Z, Zhou J, Tang J, Yang C, Chen C, et al. Influence of environmental and biological macromolecules on aggregation kinetics of nanoplastics in aquatic systems. Water Research. 2020;**186**:116316

[26] Sun H, Jiao R, Yu J, Wang D. Combined effects of particle size and humic acid Corona on the aggregation

kinetics of nanoplastics in aquatic environments. Science of the Total Environment. 2023;**901**:165987

[27] Wang M, Liu X, Zhang M, Han Q, Chen B, Cao S, et al. Comparison of microplastics heteroaggregation with MoS2 and graphene oxide nanosheets: Dependence on the configuration and impacts on aquatic transport. Journal of Hazardous Materials. 2025;**486**:137063

[28] Mao Y, Li H, Huangfu X, Liu Y, He Q. Nanoplastics display strong stability in aqueous environments: Insights from aggregation behaviour and theoretical calculations. Environmental Pollution. 2020;**258**:113760

[29] Lee C-H, Fang JK-H. Effects of temperature and particle concentration on aggregation of nanoplastics in freshwater and seawater. Science of the Total Environment. 2022;**817**:152562

[30] Korfiatis D. Microplastics aggregation, deposition, and enhancement of contaminants transport. In: Rocha-Santos T, Costa M, Mouneyrac C, editors. Handbook of Microplastics in the Environment. Cham, Switzerland: Springer; 2020. p. 1–12

[31] Alimi OS, Farner JM, Tufenkji N. Exposure of nanoplastics to freeze-thaw leads to aggregation and reduced transport in model groundwater environments. Water Research. 2021;**189**:116533

[32] Domercq P, Praetorius A, MacLeod M. The Full Multi: An open-source framework for modelling the transport and fate of nano-and microplastics in aquatic systems. Environmental Modelling & Software. 2022;**148**:105291

[33] Liu Y, Hu Y, Yang C, Chen C, Huang W, Dang Z. Aggregation kinetics of UV irradiated nanoplastics in aquatic environments. Water Research. 2019;**163**:114870

[34] Luo D, Li C, Bai X, Shi Y, Wang R. Photoaging-induced variations in heteroaggregation of nanoplastics and suspended sediments in aquatic environments: A case study on nanopolystyrene. Water Research. 2025;**268**:122762

[35] Hanif MA, Ibrahim N, Hayazi NA, Dahalan FA, Ali UFM, Jalil AA, et al. Enhancement of microplastics and nanoplastics removal via filtration method using surface-engineered palm kernel shell biochar. Separation and Purification Technology. 2025;**360**:130596

[36] Kristanti RA, Wong WL, Darmayati Y, Hatmanti A, Wulandari NF, Sibero MT, et al. Characteristics of microplastic in commercial aquatic organisms. Tropical Aquatic and Soil Pollution 2022;**2**: 134–158

[37] Jeong C-B, Kang H-M, Lee YH, Kim M-S, Lee J-S, Seo JS, et al. Nanoplastic ingestion enhances toxicity of persistent organic pollutants (POPs) in the monogonont rotifer Brachionus koreanus via multixenobiotic resistance (MXR) disruption. Environmental Science & Technology 2018;**52**: 11411–11418

[38] Walker DI, Baker-Austin C, Smith A, Thorpe K, Bakir A, Galloway T, et al. A critical review of microbiological colonisation of Nano-and Microplastics (NMP) and Their Significance to the food chain. A Critical Review of Microbiological Colonisation of Nano and Microplastics

and Their Significance to the Food Chain. 2022:184

[39] Timilsina A, Adhikari K, Yadav AK, Joshi P, Ramena G, Bohara K. Effects of microplastics and nanoplastics in shrimp: Mechanisms of plastic particle and contaminant distribution and subsequent effects after uptake. Science of the Total Environment. 2023;**894**:164999

[40] Moyal J, Dave PH, Wu M, Karimpour S, Brar SK, Zhong H, et al. Impacts of biofilm formation on the physicochemical properties and toxicity of microplastics: A concise review. Reviews of Environmental Contamination and Toxicology 2023;**261**:8

[41] Hamann L, Werner J, Haase FJ, Thiel M, Scherwaß A, Laforsch C, et al. Retention of microplastics by biofilms and their ingestion by protists in rivers. Environmental Microbiology Reports. 2024;**16**(5):e70016

[42] Pan I, Umapathy S, Issac PK, Rahman MM, Guru A, Arockiaraj J. The bioaccessibility of adsorbed heavy metals on biofilm-coated microplastics and their implication for the progression of neurodegenerative diseases. Environmental Monitoring and Assessment. 2023;**195**(11):1264

[43] Di Natale M, Catarino AI, Summers S, Boyle D, Torri M, Nicosia A, et al. Influence of microplastic-associated biofilms on the bioavailability of a mixture of cadmium and benzo [a] pyrene by the analysis of biomarker gene expression in larval zebrafish. Ecological Indicators. 2023;**152**:110369

[44] Miao L, Hou J, You G, Liu Z, Liu S, Li T, et al. Acute effects of nanoplastics and microplastics on periphytic biofilms depending on particle size, concentration and surface modification. Environmental Pollution. 2019;**255**:113300

[45] Hao T, Gao Y, Li Z-C, Zhou -X-X, Yan B. Size-dependent uptake and depuration of nanoplastics in Tilapia (Oreochromis niloticus) and distinct intestinal impacts. Environmental Science & Technology. 2023;**57** (7):2804–2812

[46] Zhang Y, Goss GG. Potentiation of polycyclic aromatic hydrocarbon uptake in zebrafish embryos by nanoplastics. Environmental Science: Nano. 2020;**7** (6):1730–1741

[47] Setälä O, Fleming-Lehtinen V, Lehtiniemi M. Ingestion and transfer of microplastics in the planktonic food web. Environmental Pollution. 2014;**185**:77–83

[48] Cole M. The Impacts of Microplastics on Zooplankton. Exeter, United Kingdom: University of Exeter (United Kingdom); 2014

[49] Liu W, Liao H, Wei M, Junaid M, Chen G, Wang J. Biological uptake, distribution and toxicity of micro (nano) plastics in the aquatic biota: A special emphasis on size-dependent impacts. TrAC Trends in Analytical Chemistry. 2024;**170**:117477

[50] Chang X, Wang W-X. Differential cellular uptake and trafficking of nanoplastics in two hemocyte subpopulations of mussels Perna viridis. Journal of Hazardous Materials. 2024;**471**:134388

[51] Chung H-E, Park D-H, Choy J-H, Choi S-J. Intracellular trafficking pathway of layered double hydroxide

nanoparticles in human cells: Size-dependent cellular delivery. Applied Clay Science. 2012;**65**:24–30

[52] Hofmann D, Tenzer S, Bannwarth MB, Messerschmidt C, Glaser S-F, Schild H, et al. Mass spectrometry and imaging analysis of nanoparticle-containing vesicles provide a mechanistic insight into cellular trafficking. ACS Nano 2014;**8**:10077–10088

[53] Gravely M, Roxbury D. Multispectral fingerprinting resolves dynamics of nanomaterial trafficking in primary endothelial cells. ACS Nano. 2021;**15**(7):12388–12404

[54] Wang T, Wang L, Li X, Hu X, Han Y, Luo Y, et al. Size-dependent regulation of intracellular trafficking of polystyrene nanoparticle-based drug-delivery systems. ACS Applied Materials & Interfaces 2017;**9**: 18619–18625

[55] Guarnieri D, Guaccio A, Fusco S, Netti PA. Effect of serum proteins on polystyrene nanoparticle uptake and intracellular trafficking in endothelial cells. Journal of Nanoparticle Research. 2011;**13**(9):4295–4309

[56] Geremia E, Muscari Tomajoli MT, Murano C, Petito A, Fasciolo G. The impact of micro-and nanoplastics on aquatic organisms: Mechanisms of oxidative stress and implications for human health—a review. Environments. 2023;**10**(9):161

[57] Zhang C, Li Y, Yu H, Ye L, Li T, Zhang X, et al. Nanoplastics promote arsenic-induced ROS accumulation, mitochondrial damage and disturbances in neurotransmitter metabolism of zebrafish (Danio rerio). Science of the Total Environment. 2023;**863**:161005

[58] Gu Y, Xu D, Liu J, Chen Y, Wang J, Song Y, et al. Bioaccumulation of functionalized polystyrene nanoplastics in sea cucumber Apostichopus japonicus (Selenka, 1867) and their toxic effects on oxidative stress, energy metabolism and mitochondrial pathway. Environmental Pollution. 2023; **319**:121015

[59] Yang S, Zhang T, Ge Y, Cheng Y, Yin L, Pu Y, et al. Ferritinophagy mediated by oxidative stress-driven mitochondrial damage is involved in the polystyrene nanoparticles-induced ferroptosis of lung injury. ACS Nano 2023;**17**:24988–25004

[60] Kumar N, Kumar S, Shukla A, Kumar S, Singh RK, Ulasov I, et al. Mitochondrial-mediated apoptosis as a therapeutic target for FNC (2′-deoxy -2′-b-fluoro-4′-azidocytidine)-induced inhibition of Dalton's lymphoma growth and proliferation. Discover Oncology 2024;**15**:16

[61] Liu X, Yang J, Li Z. Transcriptomic analysis of oxidative stress mechanisms induced by acute nanoplastic exposure in Sepia esculenta larvae. Frontiers in Physiology. 2023;**14**:1250513

[62] Sun Z, Wen Y, Zhang F, Fu Z, Yuan Y, Kuang H, et al. Exposure to nanoplastics induces mitochondrial impairment and cytomembrane destruction in Leydig cells. Ecotoxicology and Environmental Safety. 2023;**255**:114796

[63] Kumar N, Shukla A, Kumar S, Ulasov I, Singh RK, Kumar S, et al. FNC (4′-Azido-2′-deoxy-2′-fluoro (Arbino) cytidine) as an effective therapeutic agent for NHL: ROS

generation, cell cycle arrest, and mitochondrial-mediated apoptosis. Cell Biochemistry and Biophysics 2024;**82**:623–639

[64] Li Z, Feng C, Pang W, Tian C, Zhao Y. Nanoplastic-induced genotoxicity and intestinal damage in freshwater benthic clams (Corbicula fluminea): Comparison with microplastics. ACS Nano. 2021;**15** (6):9469–9481

[65] Park IH, Geum SW, Yeo M-K. Reduced cellular process and developmental process genotoxicity of polystyrene nanoplastics in zebrafish embryogenesis using Aurelia aurita proteins. Molecular & Cellular Toxicology. 2023;**19**(4):829–842

[66] Shyamasundar S, Ng CT, Lanry Yung LY, Dheen ST, Bay BH. Epigenetic mechanisms in nanomaterial-induced toxicity. Epigenomics. 2015;7(3):395–411

[67] Gong H, Li R, Li F, Guo X, Xu L, Gan L, et al. Toxicity of nanoplastics to aquatic organisms: Genotoxicity, cytotoxicity, individual level and beyond individual level. Journal of Hazardous Materials. 2023;**443**:130266

[68] Patil NA, Gade W, Deobagkar DD. Epigenetic modulation upon exposure of lung fibroblasts to TiO2 and ZnO nanoparticles: Alterations in DNA methylation. International Journal of Nanomedicine. 2016;**11**:4509–4519

[69] Sarasamma S, Audira G, Siregar P, Malhotra N, Lai Y-H, Liang S-T, et al. Nanoplastics cause neurobehavioral impairments, reproductive and oxidative damages, and biomarker responses in zebrafish: Throwing up alarms of wide spread health risk of exposure. International Journal of Molecular Sciences 2020;**21**:1410

[70] Lei L, Liu M, Song Y, Lu S, Hu J, Cao C, et al. Polystyrene (nano) microplastics cause size-dependent neurotoxicity, oxidative damage and other adverse effects in Caenorhabditis elegans. Environmental Science: Nano 2018;**5**:2009–2020

[71] Wu B, Yu H, Yi J, Lei P, He J, Ruan J, et al. Behavioral studies of zebrafish reveal a new perspective on the reproductive toxicity of micro-and nanoplastics. Toxics 2024;**12**:178

[72] Yang H, Ju J, Wang Y, Zhu Z, Lu W, Zhang Y. Micro-and nano-plastics induce kidney damage and suppression of innate immune function in zebrafish (Danio rerio) larvae. Science of the Total Environment. 2024;**931**:172952

[73] Marino R, Melillo D, Italiani P, Boraschi D. Environmental stress and nanoplastics' effects on Ciona robusta: Regulation of immune/stress-related genes and induction of innate memory in pharynx and gut. Frontiers in Immunology. 2023;**14**:1176982

[74] Cole M, Liddle C, Consolandi G, Drago C, Hird C, Lindeque PK, et al. Microplastics, microfibres and nanoplastics cause variable sub-lethal responses in mussels (Mytilus spp.). Marine Pollution Bulletin. 2020;**160**:111552

[75] Zhou C, Wang Z, Ran M, Liu Y, Song Z. Nano-selenium ameliorates microplastics-induced injury: Histology, antioxidant capacity, immunity and intestinal microbiota of grass carp (Ctenopharyngodon idella). Ecotoxicology and Environmental Safety. 2024;**285**:117128

[76] Kumar N, Delu V, Ulasov I, Kumar S, Singh RK, Kumar S, et al.

Pharmacological insights: Mitochondrial ROS generation by FNC (Azvudine) in Dalton's lymphoma cells revealed by super resolution imaging. Cell Biochemistry and Biophysics 2024;**82**:873–883

[77] Brandts I, Garcia-Ordoñez M, Tort L, Teles M, Roher N. Polystyrene nanoplastics accumulate in ZFL cell lysosomes and in zebrafish larvae after acute exposure, inducing a synergistic immune response in vitro without affecting larval survival in vivo. Environmental Science: Nano. 2020; **7**(8):2410–2422

[78] Wang F, Salvati A, Boya P. Lysosome-dependent cell death and deregulated autophagy induced by amine-modified polystyrene nanoparticles. Open Biology. 2018; **8**(4):170271

[79] Xuan L, Wang Y, Qu C, Yi W, Yang J, Pan H, et al. Exposure to polystyrene nanoplastics induces abnormal activation of innate immunity via the cGAS-STING pathway. Ecotoxicology and Environmental Safety. 2024;**275**:116255

[80] Zhang C, Li L, Alava JJ, Yan Z, Chen P, Gul Y, et al. Female zebrafish (Danio rerio) exposure to polystyrene nanoplastics induces reproductive toxicity in mother and their offspring. Aquatic Toxicology. 2024;**273**:107023

[81] Gao L, Xiong X, Chen C, Luo P, Li J, Gao X, et al. The male reproductive toxicity after nanoplastics and microplastics exposure: Sperm quality and changes of different cells in testis. Ecotoxicology and Environmental Safety. 2023;**267**:115618

[82] Yi J, Ma Y, Ruan J, You S, Ma J, Yu H, et al. The invisible Threat: Assessing the reproductive and transgenerational impacts of micro-and nanoplastics on fish. Environment International. 2024;**183**:108432

[83] He Y, Yin R. The reproductive and transgenerational toxicity of microplastics and nanoplastics: A threat to mammalian fertility in both sexes. Journal of Applied Toxicology. 2024; **44**(1):66–85

[84] Zhao Q, Fang Z, Wang P, Qian Z, Yang Y, Ran L, et al. Polylactic acid micro/nanoplastic exposure induces male reproductive toxicity by disrupting spermatogenesis and mitochondrial dysfunction in mice. ACS Nano 2025;**19**:5589–5603

[85] Chen H, Gu Y, Jiang Y, Yu J, Chen C, Shi C, et al. Photoaged polystyrene nanoplastics result in transgenerational reproductive toxicity associated with the methylation of histone H3K4 and H3K9 in Caenorhabditis elegans. Environmental Science & Technology 2023; **57**:19341–19351

[86] Wade MJ, Bucci K, Rochman CM, Meek MH. Microplastic exposure is associated with epigenomic effects in the model organism Pimephales promelas (fathead minnow). Journal of Heredity. 2025;**116**(2):113–125

[87] Zeng M, Zhan C, Li Y, Liao H, Liu W, Chen G, et al. Melatonin prevents the transgenerational toxicity of nanoplastics in zebrafish (Danio rerio). Science of the Total Environment. 2024;**953**:176043

[88] Ren X, Liu Z, Zhang R, Shao Y, Duan X, Sun B, et al. Nanoplastics aggravated TDCIPP-induced transgenerational developmental neurotoxicity in zebrafish depending on

the involvement of the dopamine signaling pathway. Environmental Toxicology and Pharmacology. 2024;**108**:104436

[89] Junaid M, Liu S, Chen G, Liao H, Wang J. Transgenerational impacts of micro (nano) plastics in the aquatic and terrestrial environment. Journal of Hazardous Materials. 2023;**443**:130274

[90] Yadav DK, Samantaray BP, Kumar R. Effect of alternative natural diet on microplastic ingestion, functional responses and trophic transfer in a tri-trophic coastal pelagic food web. Science of the Total Environment. 2024;**950**:174999

[91] Jiménez-Contreras J, Fernández-Medina RI, Fernández-Araiza MA. Microplastics pollution in tropical lakes: Water, zooplankton, and fish in Central Mexico. Environmental Monitoring and Assessment. 2024;**196**(9):813

[92] Akkajit P, Yuankerd S, Sukkuea A. Exploring microplastics in seawater and zooplankton on the Eastern Coast of Thailand: A case study in Phuket Province. Applied Environmental Research. 2024;**46**(1):007

[93] Cátia G, Marta M, Paula S, Maria Helena C. Ingestion of microplastics by Mediterranean mussel (Mytilus galloprovincialis): Influence of particle size on histopathological and oxidative stress responses. Frontiers in Marine Science. 2016;**3**:80

[94] Stienbarger CD, Joseph J, Athey SN, Monteleone B, Andrady AL, Watanabe WO, et al. Direct ingestion, trophic transfer, and physiological effects of microplastics in the early life stages of Centropristis striata, a commercially and recreationally valuable fishery species. Environmental Pollution. 2021;**285**:117653

[95] Meng L, Zheng S, Zhao Y, Liu T, Liang J, Zhu M, et al. Seasonal microplastic ingestion by carnivorous chaetognaths in Jiaozhou Bay, China: Field evidence revealing microplastic trophic transfer. Journal of Hazardous Materials. 2024;**478**:135532

[96] Susanti N, Mardiastuti A, Wardiatno Y, editors. Microplastics and the impact of plastic on wildlife: A literature review. IOP Conference Series: Earth and Environmental Science; 2020: IOP Publishing

[97] Shu R, Hu W, Gao S, Zhang S, Li Z, Liang B, et al. Transfer pattern of microplastics at an individual level: A case study of two typical Sciaenidae fish in coastal waters. Science of the Total Environment. 2023;**901**:165570

[98] Pei Y, Lin Y, Guo J, Luo K, Wu J, et al. Microplastics in wild fish in the Three Gorges Reservoir, China: A detailed investigation of their occurrence, characteristics, biomagnification and risk. Journal of Hazardous Materials. 2024;**478**:135551

[99] Miller ME, Motti CA, Hamann M, Kroon FJ. Assessment of microplastic bioconcentration, bioaccumulation and biomagnification in a simple coral reef food web. Science of the Total Environment. 2023;**858**:159615

[100] Covernton GA, Cox KD, Fleming WL, Buirs BM, Davies HL, Juanes F, et al. Large size (> 100-μm) microplastics are not biomagnifying in coastal marine food webs of British Columbia, Canada. Ecological Applications. 2022;**32**(7):e2654

[101] Alava JJ. Modeling the bioaccumulation and biomagnification potential of microplastics in a cetacean foodweb of the northeastern pacific: A prospective tool to assess the risk exposure to plastic particles. Frontiers in Marine Science. 2020;7:566101

[102] Berlino M, Sarà G, Mangano M. Functional trait-based evidence of microplastic effects on aquatic species. Biology. 2023;**12**(6):811

[103] Yang Y, Guo Y, O'Brien AM, Lins TF, Rochman CM, Sinton D. Biological responses to climate change and nanoplastics are altered in concert: Full-factor screening reveals effects of multiple stressors on primary producers. Environmental Science & Technology. 2020;**54**(4):2401–2410

[104] Montoya D, Rastelli E, Casotti R, Manna V, Trano AC, Balestra C, et al. Microplastics alter the functioning of marine microbial ecosystems. Ecology and Evolution 2024;**14**:e70041

[105] Zhou Y, He G, Bhagwat G, Palanisami T, Yang Y, Liu W, et al. Nanoplastics alter ecosystem multifunctionality and may increase global warming potential. Global Change Biology 2023;**29**:3895–3909

[106] Murphy-Hagan C, Gray AB, Singh S, Hapich H, Cowger W, Seeley ME, et al. Microplastic characterization and transport mode—A flow-integrated approach to sampling urban waterways. Environmental Research. 2025;**269**:120908

[107] Cardoso-Mohedano JG, Ruiz-Fernández AC, Sanchez-Cabeza J-A, Camacho-Torres SM, Ontiveros-Cuadras JF. Microplastics transport in a low-inflow estuary at the entrance of the Gulf of California. Science of the Total Environment. 2023;**870**:161825

[108] Xu Q, Xing R, Sun M, Gao Y, An L. Microplastics in sediments from an interconnected river-estuary region. Science of the Total Environment. 2020;**729**:139025

[109] He B, Smith M, Egodawatta P, Ayoko GA, Rintoul L, Goonetilleke A. Dispersal and transport of microplastics in river sediments. Environmental Pollution. 2021;**279**:116884

[110] Chen H, Cheng Y, Wang Y, Ding Y, Wang C, Feng X, et al. Microplastics: A potential proxy for tracing extreme flood events in estuarine environments. Science of the Total Environment. 2024;**918**:170554

[111] Unice K, Weeber M, Abramson M, Reid R, Van Gils J, Markus A, et al. Characterizing export of land-based microplastics to the estuary-Part I: Application of integrated geospatial microplastic transport models to assess tire and road wear particles in the Seine watershed. Science of the Total Environment. 2019;**646**:1639–1649

[112] van Wijnen J, Ragas AM, Kroeze C. Modelling global river export of microplastics to the marine environment: Sources and future trends. Science of the Total Environment. 2019;**673**:392–401

[113] Yang M, Wang W-X. Differential cascading cellular and subcellular toxicity induced by two sizes of nanoplastics. Science of the Total Environment. 2022;**829**:154593

[114] Contino M, Ferruggia G, Indelicato S, Pecoraro R, Scalisi EM,

Salvaggio A, et al. Sublethal effects of polystyrene nanoplastics on the embryonic development of Artemia salina (Linnaeus, 1758). Animals 2023;**13**:3152

[115] Wang B, Junaid M, Imran M, Wei L, Chen G, Wang J. Environmental and sublethal concentrations of polystyrene nanoplastics induced antioxidant system, transcriptomic responses, and disturbed gut microbiota in Oyster Magallana Hongkongensis. Journal of Agricultural and Food Chemistry. 2024;**72**(24):13581–13592

[116] Fang C, Luo Y, Naidu R. Microplastics and nanoplastics analysis: Options, imaging, advancements and challenges. TrAC Trends in Analytical Chemistry. 2023;**166**:117158

[117] Mowla M, Shakiba S, Louie SM. Selective quantification of nanoplastics in environmental matrices by asymmetric flow field-flow fractionation with total organic carbon detection. Chemical Communications. 2021;**57**(96):12940–12943

[118] Schwaferts C, Sogne V, Welz R, Meier F, Klein T, Niessner R, et al. Nanoplastic analysis by online coupling of Raman microscopy and field-flow fractionation enabled by optical tweezers. Analytical Chemistry 2020;**92**:5813–5820

[119] Eliso MC, Billè B, Cappello T, Maisano M. Polystyrene micro-and nanoplastics (PS MNPs): A review of recent advances in the use of-omics in PS MNP toxicity studies on aquatic organisms. Fishes. 2024;**9**(3):98

[120] Helal M, Liu M, Chen H, Fang M, Qiu W, Kjeldsen F, et al. Multi-omics in nanoplastic research: A spotlight on aquatic life. Frontiers of Environmental Science & Engineering 2024;**18**:133

[121] Bhagat J, Zang L, Nishimura N, Shimada Y. Application of omics approaches for assessing microplastic and nanoplastic toxicity in fish and seafood species. TrAC Trends in Analytical Chemistry. 2022;**154**:116674

[122] Dawson AL, Bose U, Ni D, Nelis JLD. Unravelling protein Corona formation on pristine and leached microplastics. Microplastics and Nanoplastics. 2024;**4**(1):9

[123] Gao L, Su Y, Mehmood T, Bao R, Peng L. Microplastics leachate may play a more important role than microplastics in inhibiting microalga Chlorella vulgaris growth at cellular and molecular levels. Environmental Pollution. 2023;**328**:121643

[124] Liu C, Zong C, Chen S, Chu J, Yang Y, Pan Y, et al. Machine learning-driven QSAR models for predicting the cytotoxicity of five common microplastics. Toxicology. 2024;**508**:153918

[125] Nelis JL, Schacht VJ, Dawson AL, Bose U, Tsagkaris AS, Dvorakova D, et al. The measurement of food safety and security risks associated with micro-and nanoplastic pollution. TrAC Trends in Analytical Chemistry. 2023;**161**:116993

[126] Basumatary T, Biswas D, Boro S, Nava AR, Narayan M, Sarma H. Dynamics and impacts of Microplastics (MPs) and Nanoplastics (NPs) on ecosystems and biogeochemical processes: The need for robust regulatory frameworks. ACS Omega. 2025;**10**(17):17051–17069

[127] Duncan TV, Khan SA, Patri AK, Wiggins S. Regulatory science perspective on the analysis of microplastics and nanoplastics in human food. Analytical Chemistry. 2024;**96**(11):4343–4358

[128] Kokalj AJ, Kalčíková G, Selonen S, Bosker T, Drobne D, Dvořáková D, et al. Strategy towards producing relevant and reliable data for the hazard assessment of micro-and nanoplastics in agricultural soils. TrAC Trends in Analytical Chemistry. 2024;**172**:117567

[129] Xie L, Ma M, Ge Q, Liu Y, Zhang L. Machine learning advancements and strategies in microplastic and nanoplastic detection. Environmental Science & Technology. 2025;**59**(18):8885–8899

Chapter 4

Toxicity of Microplastics and Nanoplastics in Aquatic Systems in Interaction with Biological – Chemical Contaminants: Advances in Remediations

Kassim Chabi, Chengsong Ye, Mahmoud Gad, Neveen M. Rizk and Xin Yu

Abstract

Microplastics and nanoplastics (MPs/NPs) have recently been recognized as critical emerging pollutants with significant effects on aquatic ecosystems and human health. Their nanoscale to microscale dimensions, extensive surface area, and hydrophobic properties enable them to adsorb and concentrate various contaminants, including pathogenic microorganisms, heavy metals, antibiotics, and persistent organic pollutants. These contaminant–plastic interactions not only enhance the mobility and persistence of hazardous substances in aquatic environments but also intensify their ecological toxicity, facilitate bioaccumulation, and drive trophic transfer across food webs. Consequently, MPs/NPs act as both pollutants and pollutant carriers, amplifying ecological disturbances and public health risks. This chapter provides a comprehensive overview of the toxicological impacts of MPs/NPs, their environmental dynamics, and the mechanistic pathways governing their interactions with co-occurring contaminants. Special attention is given to their role as vectors of pollutants and their impacts on aquatic biota and human systems. Recent advancements in removal technologies, ranging from conventional physical treatments to novel membrane and electrochemical methods, are critically discussed. The chapter also highlights current knowledge gaps, limitations in scalability, and the need for harmonized regulations. Ultimately, it calls for integrated strategies combining innovative science, sustainable engineering, and international policy frameworks to mitigate the escalating microplastic crisis.

Keywords: microplastics, nanoplastics, ecotoxicology, co-contaminants, environmental remediation

1. Introduction

Microplastics (MPs, < 5 mm) and nanoplastics (NPs, < 1 μm) are now recognized as pervasive pollutants that permeate aquatic ecosystems, terrestrial soils, and even the atmosphere [1]. These particles come from the breakdown of larger plastic debris or are directly released through industrial processes and consumer products [2–4]. Their small size, high surface-to-volume ratio, and chemical persistence enable MPs/NPs to act as carriers for diverse environmental contaminants, including pathogenic microorganisms, heavy metals, antibiotics, and persistent organic pollutants (POPs) [5, 6]. The resulting multipollutant complexes pose heightened ecological and toxicological risks. In aquatic systems, MPs/NPs readily adsorb hydrophobic organic contaminants such as polycyclic aromatic hydrocarbons (PAHs), polychlorinated biphenyls (PCBs), and polybrominated diphenyl ethers (PBDEs), thereby modifying their environmental fate, extending their mobility, and altering bioavailability [7]. These combined pollutants can intensify toxic effects on aquatic organisms, disrupt trophic interactions, and accumulate along food webs, ultimately contributing to human exposure through dietary intake. Recent studies have begun to provide quantitative evidence of MPs/NPs ingestion, associated contaminant burdens, and their toxicological implications for human health [8].

Despite the indispensable role of plastics in modern society, ranging from medical equipment and food packaging to construction materials and electronics, their environmental persistence presents a growing sustainability challenge [9–11]. Most plastics are resistant to degradation and poorly recycled; only a minor fraction is recovered, while the majority accumulates in the environment and fragments into MPs/NPs over time. Although polymers such as polyethylene (PE), polypropylene (PP), polystyrene (PS), and PE terephthalate (PET) provide essential utility, their transformation into MPs/NPs poses significant ecological and health risks if unmanaged [12].

Understanding the mechanisms of toxicity associated with MPs/NPs, particularly when acting in concert with biological and chemical co-contaminants, is crucial for assessing their full impact. Interactions with microbial, antibiotic residues, heavy metals, and endocrine-disrupting compounds may amplify environmental stress and contribute to antimicrobial resistance, ecosystem imbalance, and human health risks. However, the synergistic or antagonistic effects of these interactions remain poorly characterized, particularly under environmentally relevant conditions.

This chapter aims to (1) explore the composition, environmental behavior, and toxicological mechanisms of MPs/NPs; (2) assess their interactions with co-contaminants such as pathogens, antibiotics, heavy metals, and POPs; and (3) evaluate current advances and limitations in remediation technologies. In doing so, it highlights the urgent need for interdisciplinary approaches that bridge environmental science, toxicology, microbiology, and engineering to mitigate the escalating risks posed by MPs/NPs pollution.

2. Environmental degradation of plastics into MPs/NPs and their fate

Plastics exhibit exceptional chemical stability and resistance to degradation, enabling their persistence in natural environments over extended timescales. Once released into aquatic or terrestrial systems, larger plastic debris gradually fragments into MPs and NPs through a combination of physical, chemical, and biological processes. The rate and extent of this degradation are modulated by environmental conditions such as mechanical abrasion, ultraviolet (UV) radiation, temperature fluctuations, microbial activity, and interactions with natural organic matter or soil minerals [13, 14]. Photodegradation driven by solar UV exposure promotes polymer chain scission, resulting in surface cracking and progressive fragmentation. Microbial degradation, facilitated by bacterial and fungal enzymatic activity, further contributes to polymer breakdown, particularly in weathered or surface-aged plastics. Environmental aging also induces profound morphological and physicochemical changes, including increased surface roughness, porosity, and the formation of oxygen-containing functional groups. These transformations not only accelerate fragmentation but also enhance the sorption affinity and chemical reactivity of MPs and NPs, thereby amplifying their role as vectors for environmental contaminants. Aquatic systems act as major sinks and transport vectors for MPs/NPs. Once released into freshwater or marine environments, MPs/NPs are subject to buoyancy-driven transport, biofouling-induced sedimentation, and redistribution by hydrodynamic forces. These processes influence their vertical and horizontal distribution, increasing their likelihood of being ingested by aquatic organisms. Bioaccumulation through trophic transfer has been documented across multiple taxa, including plankton, fish, and bivalves, ultimately linking environmental MPs/NPs exposure to human dietary intake [15]. Understanding the environmental transformation of plastics into MPs/NPs and their physicochemical evolution in natural systems is essential to assess their persistence, ecological impact, and risk to food safety and public health. The degradation mechanisms and environmental influencing factors of plastics are summarized in **Table 1**.

Degradation process	Mechanisms	Key environmental influencers	Resulting changes	References
Photodegradation	UV radiation induces polymer chain scission and oxidation	Solar exposure, wavelength intensity, temperature	Surface cracking, formation of smaller particles, and introduction of oxygen-containing functional groups	[16, 17]
Mechanical degradation	Physical abrasion causes fragmentation	Wave action, sediment transport, turbidity, particle collisions	Surface area, size reduction, and irregular morphology increased	[18, 19]
Chemical degradation	Oxidation, hydrolysis, or acid–base reactions can alter the polymer structure	pH, salinity, redox potential, and presence of oxidants or pollutants	Loss of polymer integrity, increased hydrophilicity, enhanced leaching of additives	[20, 21]
Biodegradation	Microorganisms enzymatically cleave polymer chains	Microbial community composition, biofilm development, and nutrient availability	Polymer mineralization, formation of oligomers or CO_2, and partial breakdown products	[22, 23]
Thermal degradation	Elevated temperatures promote molecular rearrangement or decomposition	Ambient temperature, climate region, and burning events	Accelerated fragmentation, release of toxic byproducts	[21]

Table 1.
Degradation mechanisms of plastics and key environmental influencers.

3. Composition and toxicity of plastics

3.1 Composition of plastics

Plastic materials are composed not only of base polymers such as PE, PP, PS, polyvinyl chloride (PVC), and PET, but also a complex suite of chemical additives that modify their mechanical, thermal, and optical properties [11, 13]. These additives, incorporated during polymer processing, include plasticizers, flame retardants, stabilizers, antioxidants, pigments, and surfactants, many of which are not chemically bound to the polymer matrix and are prone to leaching under environmental conditions [24]. Several of these compounds, such as bisphenol A, phthalates, and alkylphenols, are recognized for their toxicity, endocrine-disrupting potential, and ecological persistence. Furthermore, emerging forms of functionalized NPs, such as surface-modified PS–COOH or PS–NH_2, exhibit altered reactivity that can enhance sorption affinity for toxicants and increase cellular uptake in both aquatic and mammalian systems [25]. The diversity in polymer types and additive compositions introduces considerable variability in the environmental behavior, sorption dynamics, and toxicological profiles of MPs/NPs. Therefore, characterizing the full chemical makeup of plastic debris is critical for risk assessment, especially when evaluating synergistic interactions with co-contaminants in aquatic environments. Toxic compounds in plastics are listed in **Table 2**.

Category	Representative additives	Known hazards
Plasticizers	Diisooctyl phthalate, benzyl butyl phthalate, bis (2-ethylhexyl) phthalate, dibutyl phthalate, diheptyl adipate, diethyl phthalate, diisobutyl phthalate, tris(2-chloroethyl) phosphate	Endocrine disruption, reproductive toxicity, developmental toxicity
Flame retardants	Polybrominated diphenyl ethers (PBDEs), decabromodiphenylethane, tetrabromobisphenol A, tris(2-chloroethyl) phosphate, hexabromocyclohexane, boric acid	Persistent organic pollutants (POPs) are associated with neurotoxicity and carcinogenicity
Stabilizers and antioxidants	Bisphenol A, lead and cadmium compounds, nonylphenol, octylphenol, butylated hydroxytoluene (BHT), Irganox 1010, Cyanox 2246, tris-nonylphenyl phosphate (TNPP), tris (2,4-di-tert-butylphenyl) phosphite	Endocrine disruption, liver toxicity, and bioaccumulation
Slip agents	Erucamide, oleamide, fatty acid esters, zinc stearate, synthetic waxes	Environmental persistence, surface bioaccumulation
Curing agents	Diaminodiphenylmethane, dichloromethylenedianiline, formaldehyde-aniline resins, hydrazine, triazine epoxides (TGIC)	Respiratory sensitization, mutagenicity, suspected carcinogenicity
Pigments and biocides	Cobalt diacetate, cadmium, chromium (VI), lead, arsenic, organotin compounds, triclosan	Aquatic toxicity, genotoxicity, and endocrine disruption
Hazard summary	–	Carcinogenicity, endocrine disruption, aquatic toxicity, persistence in electronic and construction waste

Information reported and sources: Study by Hahladakis et al. [11].

Table 2.
Representative chemical additives in plastics and associated hazards.

3.2 Mechanisms of toxicity transmission from plastics

MPs/NPs possess large surface-area-to-volume ratios and high hydrophobicity, characteristics that make them effective sorbents for a wide spectrum of environmental contaminants, including heavy metals, organic pollutants, and microbial pathogens. These properties enable MPs/NPs to act as environmental vectors, facilitating the transport and accumulation of hazardous substances across ecosystems. The co-occurrence of plastics with other pollutants often leads to altered environmental behavior, increased persistence, and enhanced bioavailability of the associated toxicants. Consequently, these interactions can intensify ecological risks by promoting the uptake of pollutants in aquatic organisms and contributing to bioaccumulation and trophic transfer. In particular, the synergistic toxicity of MPs/NPs with adsorbed contaminants has been implicated in disrupting aquatic food webs, spreading antimicrobial resistance, and promoting the proliferation of pathogenic microorganisms in water bodies. These effects are not only environmentally consequential but also pose direct and indirect threats to public health. Experimental studies have further demonstrated that the toxicological impact of MPs/NPs is time-dependent, with cellular damage increasing with exposure duration. For instance, *in vitro* assays using human colon epithelial cells exposed to a fixed concentration of MPs revealed progressive cytotoxicity: 18.4% after 6 hours, 24.9% after 24 hours, and 42.8% after 48 hours, indicating that prolonged exposure exacerbates cellular injury and stress responses [26]. Such observations demonstrate the complex and multifactorial nature of plastic-induced toxicity, which arises not only from the polymers themselves but also from their capacity to act as carriers of environmental contaminants.

4. Interaction between MPs/NPs and co-contaminants

4.1 Interaction of MPs/NPs with microorganisms

Interactions between MPs/NPs and microbial communities are increasingly recognized as ecologically significant, particularly in aquatic ecosystems. Studies have documented key interaction pathways, including microbial adhesion, biofilm formation, and the microbial degradation of plastic surfaces [27]. MPs/NPs can adsorb onto microbial cell membranes, thereby influencing microbial physiology, metabolic activity, and community structure. In plastic-contaminated waters, the accumulation of organic pollutants on MPs/NPs further exposes microorganisms to toxic compounds, disrupting the ecological balance by altering microbial diversity, composition, and functional profiles [28]. MPs/NPs also act as vectors for microbial dispersal, facilitating the transport of a wide range of pathogenic microorganisms, including bacteria such as *Pseudomonas*, *Escherichia coli*, *Salmonella*, and *Vibrio cholerae*; viruses including Rotavirus, Norovirus, Hepatitis A virus, and Adenovirus; fungi such as *Aspergillus spp.*, *Penicillium spp.*, and *Cladosporium spp.*; and protozoan pathogens including *Giardia*, *Entamoeba histolytica*, and *Cryptosporidium spp* [29–32]. These pathogens, when bound to MPs/NPs, may bypass environmental filters and water treatment barriers, posing potential risks of infection upon human exposure. The dynamic nature of microbial adsorption and desorption on plastic surfaces presents further challenges for drinking water treatment systems, particularly during disinfection stages, where microbial persistence may be enhanced. MPs/NPs provide favorable microenvironments for microbial colonization, supporting biofilm development, microbial migration, and even horizontal gene transfer among bacteria. Such processes are particularly concerning for the dissemination of antibiotic resistance genes in aquatic environments [29].

Importantly, microorganisms are not only affected by MPs/NPs but also contribute actively to their transformation. Specific bacterial and fungal taxa have demonstrated the ability to utilize plastic polymers as carbon and energy sources, initiating the biodegradation of plastic particles. This process typically involves the enzymatic cleavage of polymer chains into oligomers or monomers, followed by assimilation and mineralization. Two key classes of enzymes are involved: extracellular depolymerases, which break down polymers into smaller absorbable units, and intracellular enzymes, which further metabolize the resulting fragments within the microbial cell [30]. Moreover, microbial metabolism can produce by-products that accelerate chemical deterioration of the plastic matrix, enhancing overall degradation rates. Several microbial species, including *Pseudomonas spp.*, *Rhodococcus spp.*, *Actinomycetes*, and fungal genera such as *Penicillium* and *Aspergillus*, have been identified as efficient degraders of synthetic polymers [31]. These organisms can form dense biofilms on plastic surfaces, reinforcing both their metabolic interactions and ecological significance in plastic-contaminated habitats.

4.2 Interaction between MPs/NPs and antibiotics

The co-occurrence of MPs/NPs with antibiotic residues in aquatic environments has emerged as a critical environmental and public health concern. Owing to their large specific surface area, hydrophobicity, and small pore sizes, MPs/NPs serve as effective adsorbents for various micropollutants, including antibiotics. These interactions facilitate the persistence, transport, and bioavailability of antibiotics in water bodies, with implications for the health of aquatic ecosystems and the proliferation of antimicrobial resistance [32]. Although the exact physicochemical mechanisms underlying antibiotic sorption onto plastic surfaces remain under investigation, several contributing factors have been identified. One is biofilm formation on MPs/NPs, which can significantly enhance antibiotic retention due to the complex matrix of extracellular polymeric substances and microbial biomass that trap antibiotic molecules [33]. Moreover, MPs/NPs undergo environmentally induced transformations, such as aging, photooxidation, and surface oxidation, which modify their surface chemistry, charge, and porosity. These changes can either enhance or diminish their sorption capacity, thereby altering pollutant transport dynamics and toxicity profiles across aquatic systems (see **Figure 1**).

Combined exposure to MPs/NPs and elevated concentrations of antibiotics or disinfectants can further complicate environmental risk scenarios. Such interactions may promote the formation of secondary transformation products, including reactive oxygen species or plastic–pollutant adducts, which can intensify toxicological outcomes [34, 35]. Experimental studies have shown that co-exposure often results in additive, antagonistic, or synergistic effects, depending on the pollutant type, concentration, and organismal sensitivity. These effects include heightened physiological stress, tissue inflammation, and disruption of gut and liver function in aquatic vertebrates and invertebrates. Importantly, such interactions may also affect human health through the food chain or waterborne exposure.

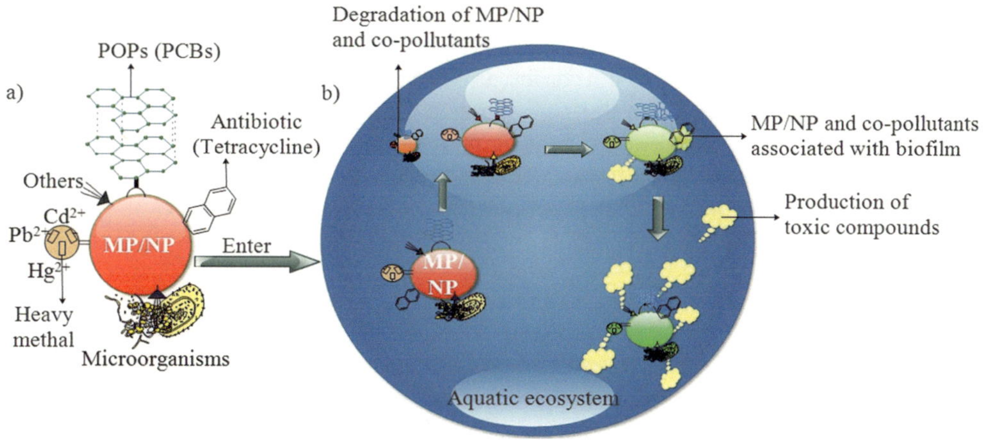

Figure 1.
Showing potential interactions between MPs/NPs and pollutants in the ecosystem. Own elaboration based on the study by Wang et al. [41]. (a) Adsorption of contaminants onto MPs/NPs. (b) Potential synergy between MPs/NPs and co-pollutants, leading to the production of toxic compounds in the aquatic ecosystem.

Supporting this concern, a study examining the interaction between polyamide 6 (PA6) MPs and the antibiotic sulfamethoxazole (SMX) reported a maximum adsorption capacity of 0.089 mg SMX per gram of virgin or aged PA6. In reservoir water, desorption from virgin PA6 reached 0.15 mg/L, confirming the role of MPs as reservoirs and secondary sources of antibiotics in aquatic systems [36].

4.3 Interaction between MPs/NPs and heavy metals

The interactions between MPs/NPs and heavy metals are of increasing concern due to their potential to alter the environmental fate, mobility, and toxicity of metallic contaminants. Numerous studies have demonstrated that MPs/NPs can effectively adsorb heavy metal ions from aquatic environments, thereby enhancing their persistence, transport capacity, and bioavailability [37]. This adsorption behavior is largely driven by the physicochemical characteristics of plastic surfaces, including large surface area, surface charge, roughness, and the presence of functional groups (e.g., carbonyl, hydroxyl, or ester groups) that facilitate metal ion binding through electrostatic attraction or complexation. Empirical evidence confirms that several environmentally relevant heavy metals, including lead (Pb^{2+}), nickel (Ni^{2+}), copper (Cu^{2+}), zinc (Zn^{2+}), and cadmium (Cd^{2+}), can be adsorbed onto both virgin and weathered plastic surfaces. For example, PET MPs have been shown to exhibit significant sorption capacity for Cu^{2+} and Zn^{2+} ions in aquatic systems, with aged PET often displaying greater affinity due to oxidative surface modification [38]. Advanced analytical techniques, such as scanning electron microscopy (SEM) combined with energy-dispersive X-ray spectroscopy (EDX), have provided visual and elemental confirmation of metal ion attachment to plastic particles, offering compelling evidence for their role as carriers of metallic contaminants [39]. The biological and ecological implications of such interactions are increasingly evident. Once ingested by aquatic organisms, metal-contaminated MPs/NPs can induce substantial toxicological effects, including oxidative stress, metabolic disruption, and impaired physiological function. A study involving the marine ciliate *Euplotes vannus* demonstrated that co-exposure to PS MPs and cadmium (Cd^{2+}) led to severe outcomes, including elevated oxidative damage, reduced carbon biomass, DNA damage, protein oxidation, and inhibited growth and reproduction [40]. These findings highlight the synergistic toxicity that arises from plastic-metal complexes and underscore the need for integrated assessments of plastic pollution that account for co-contaminant dynamics in risk evaluation frameworks.

4.4 Interaction of MPs/NPs with POPs

POPs are a group of toxic, long-lived, and bioaccumulative chemicals that include PAHs, PCBs, PBDEs, and dichlorodiphenyltrichloroethane (DDT). These compounds are frequently detected in association with MPs/NPs in various environmental matrices, particularly in aquatic ecosystems [41]. MPs/NPs can interact with POPs primarily through sorption and desorption processes, thereby influencing their environmental distribution, persistence, and ecological impact. The sorption of POPs onto plastic particles can be facilitated by the unique physicochemical properties of MPs/NPs, including a large specific surface area, hydrophobicity, the presence of surface functional groups, and electrostatic interactions. These attributes enhance the affinity of plastic surfaces for hydrophobic organic contaminants, allowing MPs/NPs to act as mobile reservoirs and vectors for POPs in aquatic environments. As these contaminated particles are transported across different habitats, they may release the adsorbed POPs under changing physicochemical conditions, such as variations in pH, salinity, or temperature.

The combined presence of MPs/NPs and POPs has been shown to result in complex toxicological interactions. These include additive, synergistic, or antagonistic effects, depending on several factors such as pollutant type, concentration, duration of exposure, and species-specific sensitivity [42]. Synergistic interactions, in particular, can amplify ecological risks by increasing pollutant bioavailability and enhancing toxicity toward aquatic organisms, as well as potentially humans [5]. For example, co-exposure may lead to elevated oxidative stress, immune dysfunction, or reproductive toxicity that surpasses the effects observed with either pollutant alone. These multi-pollutant interactions present substantial challenges for environmental risk assessment and regulatory management, as they complicate predictions of contaminant fate and toxicity. Comprehensive studies that integrate the behavior of plastic-associated pollutants, mixture toxicology, and real-world exposure scenarios are needed to inform environmental monitoring and policy development.

5. Human and ecological health concerns related to MPs/NPs exposure

The widespread production, usage, and environmental mismanagement of plastic materials have contributed to the pervasive presence of MPs/NPs across terrestrial, freshwater, and marine ecosystems (see **Table 3**). Their small size, persistence, and mobility allow these particles to infiltrate biological systems, where they may pose toxicological threats. Recent evidence has confirmed the presence of MPs/NPs

Parameters	Quantitative value	Sample matrix/Exposure context	References
MPs	2,649–2,857 particles/L 6,292 ± 10,521 particles/L	Mineral water brands	[43]
MPs	0–10^3 particles/L	Tap water	[44]
MPs	0 to > 10,000 particles/L	Surface water	[45]
MPs	0–6.8 × 10^6 particles /km^2	Surface water	[46]
MPs (annually discharged)	115–164 × 10^{12} particles of artificial fibrous MPs	Surface water	[47]
MPs/NPs	0–325 particles/L	Seawater	[48]
MPs	1–18,285 particles/L	Wastewater	[49]
MPs	3.67–4,180 MPs per individual	Zooplankton	[50]
MPs	0–106 MPs/mL	Marine organisms	[51]
MPs (daily intakes	0.068–0.19 MPs/(kg day) 0.089–0.25 MPs/(kg day)	Adults Children	[52]
MPs (daily intakes)	0.274 and 0.600 MPs/(kg day)	Adults and children	[53]

Table 3.
Summarizes environmental concentrations of MPs/NPs in various water sources (e.g., drinking water, seawater, wastewater) and human exposure levels, based on current literature.

in several human biological matrices, including fecal samples, blood plasma, placental tissues, and lung parenchyma, demonstrating their capacity for systemic absorption, translocation, and bioaccumulation within the human body. These findings have heightened global scientific interest in evaluating the potential health consequences associated with MP/NP exposure, particularly regarding inflammation, oxidative stress, endocrine disruption, and cytotoxicity (see **Table 4**). Parallel concerns exist in ecological contexts, where MPs/NPs may disrupt microbial communities, impair reproductive and metabolic functions in aquatic organisms, and propagate through food webs via trophic transfer. Given their documented presence in environmental media, drinking water, and food products, assessing the health risks associated with MPs/NPs has become a critical component of environmental toxicology and public health surveillance.

5.1 Mechanisms and potential risks of human exposure to MPs/NPs

MPs/NPs have been increasingly identified as contaminants in multiple environmental media, including drinking water [59–62], food products, and airborne particulates [63], establishing ingestion, dietary intake, and inhalation as primary exposure routes to humans (see **Figure 2.**). Within aquatic ecosystems, MPs/NPs enter the food web through ingestion by zooplankton and other lower trophic organisms. These particles are then transferred to planktivorous and piscivorous species and can ultimately bioaccumulate in edible aquatic organisms, thereby reaching human consumers through seafood consumption [64]. Once ingested, MPs/NPs may cross epithelial barriers and translocate into systemic circulation, a process influenced by their size, surface charge, concentration, and physicochemical properties [65]. NPs, in particular, exhibit the capacity to penetrate

Study/Endpoint	Model	Exposure (dose; duration)	Key toxicological outcome (effect size when available)	References
MPs/NPs in carotid plaques and clinical prognosis	Humans (carotid endarterectomy cohort)	MNPs (mainly PE; some PVC) detected in excised plaques; median follow-up ≈ 34 months	Major adverse cardiovascular events (stroke/death) risk = 4.53 for MPNP-positive versus MPNP-negative plaques	[54]
Systemic exposure – microplastics in human blood (initial report)	Humans (whole blood biomonitoring)	Polymer mass by Py–GC/MS; 77% of donors > LOQ; mean total plastic ≈ 1.6 µg/mL	Confirms systemic bioavailability of PET, PE, styrenics, PMMA	[55]
Bioaccumulation in human brain (post-mortem)	Humans (autopsy tissues)	Tissue MNP loads (Py–GC/MS) across organs	Brain 7–30 × higher MNP concentrations than liver/kidney; PE predominant; upward trend 2016→2024	[56]
Hepatotoxicity in immune-competent co-culture	*In vitro* (HepG2 + THP-1)	PMMA MPs (3–10 µm), up to 1 mg/mL, ≤ 72 h	Inflammation, oxidative stress, lipid-metabolism disruption; TEM confirms cellular uptake	[57]
Dose metrics and risk assessment guidance	Cross-model (review)	Emphasis on delivered dose; harmonized units (mass, number, surface area); particle characterization	Interpretation of dose–response depends on standardized metrics	[58]

Abbreviations: Py–GC/MS: pyrolysis gas chromatography–mass spectrometry; PMMA: microplastic polymethyl methacrylate); HepG2/THP-1: human liver co-culture model.

Table 4.
Compiles toxicological dose–response data across different organisms, including humans, animals, and in vitro *models, with emphasis on health endpoints such as hepatotoxicity, cardiovascular risks, and bioaccumulation.*

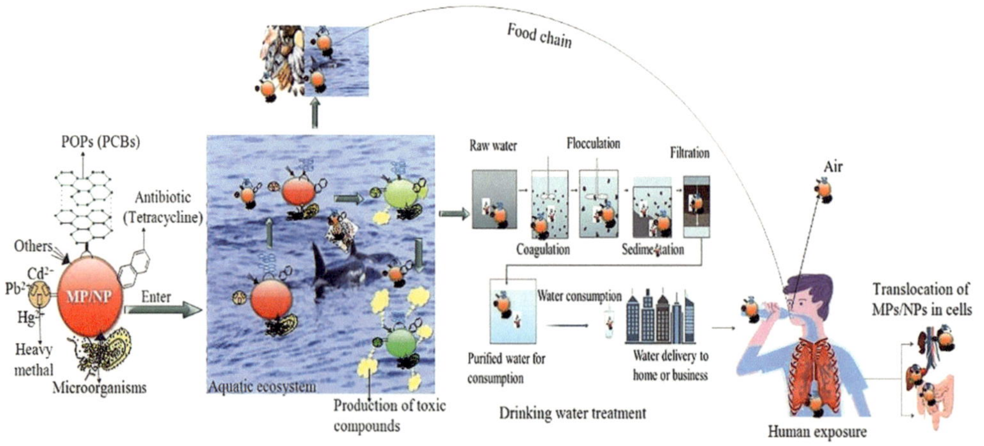

Figure 2.
Mechanism of animals and humans' exposure to MPs/NPs. Source: Study by Bakan et al. [8].

cellular membranes and accumulate in vital organs, including the liver, kidneys, and intestines [66]. Depending on their size and surface chemistry, plastic particles may persist in the body for extended periods or be eliminated via bile excretion into the intestine, urinary filtration, or fecal routes [67]. Alarming evidence of MPs/NPs in the human placenta suggests their ability to cross critical biological barriers, raising concerns about potential fetal exposure and developmental toxicity [68].

Scientific studies have provided a growing body of data on the pathways and toxicological consequences of human exposure to MPs/NPs [8]. As these particles degrade or leach additive chemicals within the body, they may release toxic compounds that exacerbate health risks. Documented adverse effects include cellular damage, oxidative stress, inflammation, and increased risk of chronic diseases. Health complications linked to MPs/NPs exposure span multiple systems, including the reproductive, cardiovascular, digestive, and respiratory systems, and are associated with serious conditions such as infertility, heart disease, and impaired gastrointestinal function [69].

In addition, MPs/NPs are increasingly recognized as endocrine-disrupting chemicals (EDCs), with established links to obesity, diabetes, breast and prostate cancers, thyroid dysfunction, cardiovascular abnormalities, cognitive delays, and neurological disorders [70, 71]. Evidence also indicates that plastic particles can induce cytotoxicity, neurotoxicity, and disruptions in hepatic and hematological functions in women, along with menstrual irregularities, tissue damage, and chronic stress responses [72]. Further studies have demonstrated adverse effects on cellular metabolism, immune function, microbial homeostasis, and even organ development, supporting concerns over their long-term biotoxicity [73]. Such findings strongly reinforce the need for stringent environmental control of MPs/NPs and emphasize the importance of comprehensive toxicological evaluation to understand and mitigate their risks to human health.

5.2 Mechanism and potential risks of animal exposure to MPs/NPs

MPs/NPs can enter aquatic environments through multiple pathways, including runoff from land surfaces, atmospheric deposition, rainfall, and direct discharge from human activities. Once introduced into surface waters, their small size and relatively low density allow them to remain suspended or float, increasing their accessibility to aquatic biota. Initially buoyant, MPs/NPs often undergo biofouling, whereby microbial colonization increases their density and promotes sedimentation. As a result, plastic particles are present throughout the water column, making them bioavailable to a wide range of organisms (see **Figure 3**). Studies have shown that NPs accumulate primarily in the gastrointestinal tracts of zooplankton species such as *Daphnia magna*, *Thamnocephalus platyurus*, and *Brachionus calyciflorus*, with ingestion localized to the gut or stomach [74]. Beyond plankton, MPs/NPs have also been detected in various fish species, bivalves [75, 76], and other aquatic invertebrates [77], confirming their pervasive uptake across trophic levels. Experimental research has further demonstrated that these plastic particles can translocate from the gut to internal tissues in organisms such as fish, clams, and *Daphnia*, indicating systemic exposure and potential bioaccumulation [66]. Moreover, MPs/NPs have been found within animal cells, and toxicological studies in mammals (e.g., mice) have linked their presence to reduced blood flow, neurological impairments, and vascular obstruction [78].

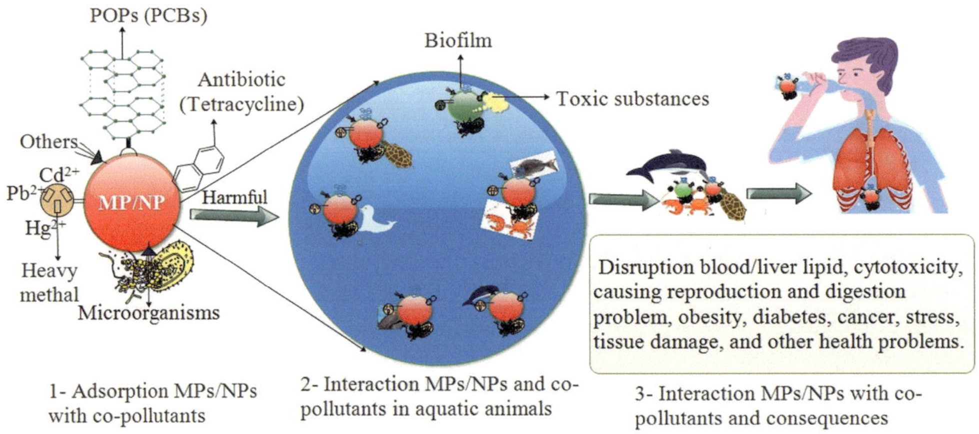

Figure 3.
MPs/NPs with co-pollutants and their potential effects [64]. The figure illustrates the adsorption and desorption of pollutants, as well as the injection of MPs/NPs with co-pollutants by aquatic animals into water, highlighting possible impacts on animal and human health.

Ecotoxicological studies have revealed both acute and chronic effects of MPs/NPs on aquatic organisms. In zooplankton, exposure to environmentally relevant concentrations of NPs for 48 hours and MPs for up to 8 days resulted in up to a 50% reduction in reproductive capacity, suggesting severe sublethal impacts even at modest exposure levels [75]. In freshwater plankton species such as *Scenedesmus obliquus*, NPs have been shown to inhibit growth and increase mortality. Similarly, exposure to MPs/NPs has induced oxidative stress, root growth inhibition, and cytogenotoxic effects in plant models such as *Allium cepa*, underscoring their potential to affect aquatic ecosystems at multiple biological levels [73]. Together, these findings demonstrate that MPs/NPs are not only bioavailable to a wide range of aquatic organisms but can also cause functional impairments, reproductive failure, and physiological toxicity. The ecological consequences of MPs/NPs are likely to be exacerbated by their interactions with co-pollutants such as heavy metals, antibiotics, and POPs, as previously detailed [7].

Although most evidence for MP/NP toxicity comes from laboratory exposures, field data indicate that risks to wild aquatic organisms are often subtler and context-dependent [79, 80]. Weight-of-evidence assessments by SAPEA and the WHO conclude that, at current environmental concentrations, widespread population-level impacts have not been demonstrated, while also emphasizing major uncertainties and the need for long-term monitoring [79, 80]. Extrapolating laboratory effects to nature is complicated because many experiments use MP doses above field levels and differ in particle types and test conditions [81, 82]. Meta-analyses across fish and invertebrates report negative but generally small and variable effect sizes for growth, reproduction, and survival, suggesting sensitivity is taxon- and context-specific [83, 84]. In wild populations, MP/NP ingestion is common and trophic transfer occurs, even to marine mammals, but clear links to health or fitness are inconsistent; several seabird studies, for example, find weak or no correlations between gut MP loads and condition, whereas others document sublethal pathology such as plastic-associated gastric fibrosis ("plasticosis") in certain species [85–88]. Overall, current field evidence points to heterogeneous, often less obvious effects than those observed under controlled, higher-dose laboratory scenarios, underscoring the need for ecologically realistic exposure studies and standardized monitoring in natural systems [81].

6. Advances in MPs/NPs removal in water treatment, challenges, and remediation approaches

The mitigation of MPs/NPs in aquatic systems has become a pressing concern for water treatment and environmental remediation. Effective removal strategies typically involve a combination of physical, chemical, and biological processes, each targeting different physicochemical properties of MPs/NPs. Among these, physical removal methods, especially those based on adsorption mechanisms, have emerged as the most widely applied and technically feasible in both drinking water and wastewater treatment systems.

The adsorption of contaminants onto MPs/NPs is governed by multiple intermolecular forces, including hydrophobic interactions, hydrogen bonding, electrostatic interactions, halogen bonding, π–π stacking, and negative charge-assisted hydrogen bonding [89]. These interactions enable MPs/NPs to act as sorbents for a variety of organic and inorganic pollutants, complicating their removal and increasing their environmental mobility. Moreover, several environmental factors, such as pH, ionic strength, aging, presence of dissolved organic matter, and mineral composition, significantly influence the adsorption capacity and behavior of MPs/NPs [90]. For instance, aged plastic particles may exhibit increased surface reactivity due to oxidation and surface roughening, enhancing their affinity for pollutant binding. As MPs/NPs are highly variable in size, density, and surface chemistry, their efficient removal requires integrated treatment approaches tailored to these properties. In subsequent sections, this chapter will explore conventional and advanced water treatment technologies, evaluate their effectiveness in removing MPs/NPs, and discuss emerging remediation strategies along with existing limitations and knowledge gaps.

6.1 Physical techniques

A range of conventional and advanced physical processes have been employed to remove MPs/NPs from water treatment systems (see **Figure 4**). These technologies primarily operate through adsorption, filtration, coagulation, and separation mechanisms, leveraging the physicochemical properties of MPs/NPs to facilitate their extraction from aqueous matrices. Adsorption-based methods, particularly those utilizing biochar and activated carbon, have shown considerable promise due to their high sorption capacity and cost-effectiveness. More recently, electrosorption employing activated carbon electrodes has gained attention as an emerging approach, offering high efficiency and potential compatibility with existing treatment infrastructures. Conventional coagulation–flocculation processes, often integrated with sedimentation or filtration, remain widely applied; in these systems, chemical coagulants promote the aggregation of MPs/NPs into larger flocs that can be effectively separated.

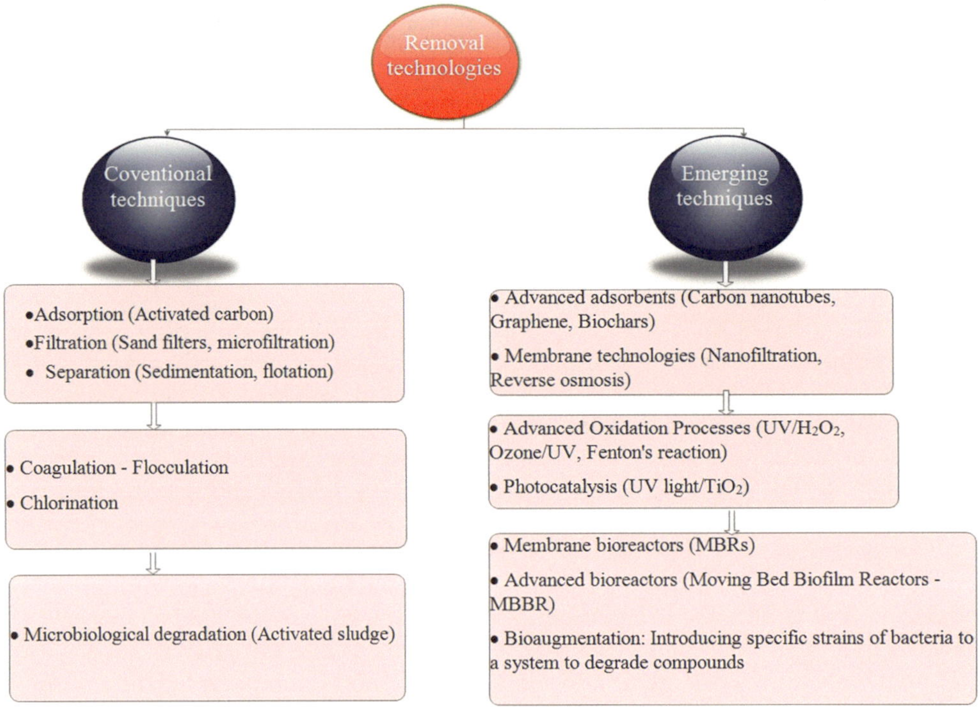

Figure 4.
Current MPs/NPs and co-contaminants removal techniques, in the water treatment processes.

In addition, advanced techniques such as membrane filtration, including ultrafiltration, nanofiltration, and reverse osmosis, as well as media-based methods like rapid sand filtration and biochar-assisted filtration, offer high removal efficiencies, particularly for nanoscale plastics that are otherwise difficult to eliminate. Magnetic-seeded filtration and metal–organic frameworks (MOFs) have also emerged as effective approaches, particularly in laboratory-scale applications. The overall removal efficiency of these physical methods depends on several factors, including particle size, surface chemistry, influent water quality, and operational parameters such as pore size and hydraulic loading. Laboratory studies have further demonstrated that the effectiveness of physical removal processes can be enhanced by designing materials with tailored surface properties that promote electrostatic attraction, hydrophobic interactions, π–π stacking, and hydrogen bonding [91]. Among physical methods, membrane and media filtration have shown particularly high performance, with removal efficiencies frequently exceeding 90%, making them suitable for full-scale applications [92]. Previous studies summarize the physical methods and efficiency of removal (see **Tables 5**, **6** and 7).

6.2 Environmental remediation approaches

The remediation of MPs/NPs from the environment remains a complex challenge due to their small size, widespread distribution, and frequent association with co-contaminants, including heavy metals, antibiotics, and pathogens. The physical and

Physical treatment	Average efficiency of MP removals (%)
Coagulation	61
Electrocoagulation	>90
Membrane bioreactor	99.9
Rapid sand filtration (RSF)	97
Disc filter	40 – 98.5
Air flotation	95
Magnetic-seeded filtration	95
Adsorption by metal–organic frameworks	90
Biochar sand filter	95

Source: Study by Subair et al. [86].

Table 5.
Physical methods with average efficiency of MP removals.

Types of plastics	Technologies	Removal efficiencies	References
MPs	Coagulation and flocculation using coagulant chemicals, sand, and activated carbon	95%	[87]
NPs	Coagulation/flocculation/sedimentation, sand/activated carbon filtration	83%	[93]
NPs	Biochar	99%	[94]
MPs/NPs	Membrane and media filtrations	99%	[95]
MPs/NP	Membrane filtration	91%–93%	[96]
MPs/NPs	Electrocoagulation process using different combinations of aluminium (Al) and stainless steel (SS) electrodes, including Al–Al, SS–SS, Al–SS, and SS–Al	94%–93.6%	[97]
Polystyrene nanoplastics (PS/NPs)	Chromium-based metal–organic framework (Cr-MOF/MIL-101).	96%	[98]

Table 6.
Studies using physical treatment methods to remove MPs/NPs in water.

chemical diversity of these particles complicates their identification, collection, and recycling at scale. Nonetheless, several strategic interventions have been proposed to mitigate MP/NP pollution and limit their release into natural ecosystems. A key preventive approach involves reducing plastic production, particularly of single-use and non-recyclable plastics, and promoting the adoption of biodegradable or compostable alternatives. On the treatment side, improving the efficiency of wastewater and stormwater treatment plants is essential to intercept MPs/NPs before effluents are discharged into surface waters. Upgrading conventional systems with advanced filtration, coagulation, or adsorption technologies can significantly enhance removal efficiency.

Captured MPs/NPs from wastewater can be sequestered in sludge during treatment. While sludge management presents its own environmental

Techniques	Specific techniques and efficiency	Pros (advantages)	Cons (limitations and challenges)	Potential for large-scale application	References
Physical	Adsorbent using (Activated Carbon/Biochar/Synthetic material) with efficiency > 90% for MP removal.	• High efficiency and capacity for targeted MPs, especially with advanced materials. • Operational simplicity, reusable (magnetic recovery), biodegradable options. • Can be tailored to specific polymer types or sizes.	• Secondary waste (spent adsorbent). • Adsorbent cost, regeneration, and disposal are significant challenges. • Efficiency drops for NPs < 20 nm. • Competitive inhibition from natural organic matter (NOM) reduces efficiency in real water.	Highly used in water treatment. Scalable production of biochar and low-cost natural adsorbents (e.g., bamboo foam).	[99–101]
	Coagulation/Sedimentation and flotation > 80%.	• High efficiency for many MPs (with optimized coagulants). • Simple integration into existing conventional water treatment plants. • Cost-effective and uses established technology. • Effective for a wide size range of MPs.	• Produces chemical sludge requiring further treatment/disposal, potentially concentrating MPs. • Efficiency is highly dependent on water chemistry (pH, organics). • Less effective for NPs and certain polymer types without tailored coagulants.	Potential for large-scale application	[102]
	Membrane filtration efficiency variable (80%–99%) for MPs/NPs Electrospun membranes: 89.9%.	• Very high removal efficiency. • Excellent for removing small MPs and even some NPs. • Reliable and well-understood technology.	• Membrane fouling increases operational energy costs and requires frequent cleaning/replacement. • High capital and operational costs for advanced membranes (NF, RO). • Concentrates MPs in a waste stream (backwash water) that must be managed.	Moderate to high; already used in advanced WWTPs, but cost and energy are barriers for widespread NP removal.	[103–105]

Techniques	Specific techniques and efficiency	Pros (advantages)	Cons (limitations and challenges)	Potential for large-scale application	References
Chemical	Coagulation using chemical coagulants (e.g., polyaluminium chloride, ferric chloride): 99.4% for MPs and < 90% for NPs.	High efficiency, simple integration into conventional water treatment.	Produces chemical sludge requiring disposal; efficiency depends on water chemistry.	Highly implemented as a primary treatment step. Sludge management is a challenge, but it presents a feasible solution for large-scale water treatment applications.	[106]
	Advanced oxidation processes (AOPs): Ozonation (with catalyst): > 96.7% for NPs Photocatalysis (e.g., TiO_2/SiO_2): 50% degradation of PMMA in seven hours.	• High energy input, costly catalysts. • Mineralizing MPs/NPs into CO_2 and H_2O. • Effective for recalcitrant polymers and can break down NPs.	• Possible destruction of MPs/NPs into harmless products, no solid waste. • Destroys MPs rather than just separating them, minimizing secondary waste. • High energy consumption leads to high operational costs. • By-product risks need addressing for full-scale.	• Moderate; pilots show promise. • Not yet optimized for large-scale MP removal; often used for disinfection.	[107]
	Electrochemical Oxidation: > 92%	• High degradation efficiency without chemical additives. • Operates at ambient conditions. • Modular and scalable system design.	• Electrode fouling and cost of durable anodes (e.g., Boron-Doped Diamond). • Energy-intensive. • Still primarily at lab-scale for MP treatment; more research needed.	The electrochemical is a feasible method for the treatment of water contaminated by MPs/NPs. Not yet optimized for large-scale.	[108, 109]

Techniques	Specific techniques and efficiency	Pros (advantages)	Cons (limitations and challenges)	Potential for large-scale application	References
Biological	Mixed cultures. Lab-scale bioreactors: Up to 98.83% removal reported. Bacteria such as *Acinetobacter, Bacillus, Pseudomonas*, and *Klebsiella* have been gradually isolated through enrichment and culture from plastic waste, sludge, and wastewater. Efficiency variable in lab-scale.	• Sustainable and utilizes wastewater nutrients, produces valuable biomass. • Low energy consumption compared to physical/chemical methods. • Environmentally friendly, and potentially mineralizes MPs.	• Very slow (requires long hydraulic retention time days/weeks), sensitive to toxicity, not effective for all polymer types (e.g., PE, PP). • MPs are concentrated in sewage sludge, which is often land-applied.	Moderate; core of WWTPs but inefficient for NPs alone. Needs bioaugmentation or longer retention times for enhanced MP removal. Not yet optimized for large-scale.	[110–112]
	Membrane bioreactors (MBR) > 98% removal for NPs when combined with sand filtration.	• Combines biological treatment and membrane filtration. • Superior effluent quality. • Compact footprint. • High and increasing adoption in advanced WWTPs, especially where water reuse is a priority.	• High capital and operational costs. • MPs and biomass exacerbate membrane fouling. • Still produces concentrated sludge requiring management.	MBR technology is the core of some the world's largest and most advanced wastewater treatment plants.	[104]
	Bioaugmentation/Enzymatic degradation	• Biodegradation towards complete mineralization. • Highly specific and can target persistent polymers of MPs/NPs. • Sustainable and green approach.	• Very slow degradation rates (months to years). • Sensitive to environmental conditions (pH, T, toxicity). • Not yet viable for large-scale application; purely at the research stage.	• Not yet viable for large-scale application.	[113]
	Bioelectrochemical: Low efficiency compared to the electrochemical method on MP/NP removal.	• Exhibit high EPS production to enhance MPs/NPs mineralization. • Aid in the reduction of the toxicity of MPs/NPs.	• Electrode bio-fouling. • Increase capital cost.	• Limited in the research conducted. • Not yet tested for large-scale application.	[114]

Table 7.
Comparison of MP/NP removal approaches: efficiency, advantages, limitations, and scalability.

considerations, emerging research suggests that with proper decontamination, MP/NP-containing sludge can be transformed for agricultural applications, such as a component of organic fertilizers. However, this practice requires strict regulatory oversight to prevent soil contamination (see: https://www.eib.org/en/essays/plastic-pollution). In parallel, source control strategies, public awareness campaigns, and producer responsibility frameworks are critical to reducing plastic inputs into the environment. Ultimately, addressing MP/NP pollution requires a multifaceted remediation approach that integrates technological innovation, policy reform, and sustainable material design.

6.3 Challenges in MPs/NPs removal

The effective removal of MPs/NPs from water systems remains technically challenging, primarily due to their minute size, diverse composition, and heterogeneous distribution. Their small dimensions allow them to bypass many conventional treatment barriers, especially during the early stages of water and wastewater processing. While biological degradation and chemical oxidation are occasionally explored as treatment options, their efficacy is limited and highly context-dependent. These methods may either partially degrade plastic polymers into less harmful fragments or, under certain conditions, contribute to the formation of secondary pollutants, thus exacerbating water quality issues [86].

In comparison, membrane filtration (e.g., ultrafiltration, nanofiltration, reverse osmosis) and media filtration (e.g., sand, activated carbon, or biochar filters) offer higher removal efficiencies. However, their performance is constrained by operational limitations such as membrane fouling [103], and adsorbent pore saturation [89], which reduces long-term efficiency and increases maintenance demands. These processes can become inefficient over time without regular cleaning, regeneration, or replacement of materials. Emerging electrochemical techniques, including electrosorption, electrocoagulation, and electrodialysis, have shown promise in removing MPs/NPs by leveraging electric fields for charge-based separation and adsorption. However, these methods are often not sufficient as standalone treatments and may require integration with conventional filtration systems to achieve comprehensive removal. Overall, the complexity of MPs/NPs removal highlights the need for multi-barrier treatment strategies, material innovation, and system optimization tailored to plastic particle characteristics, water quality parameters, and treatment objectives.

6.4 Challenges in MPs/NPs study

Monitoring and studying MPs/NPs remain fraught with significant methodological and analytical challenges. A primary obstacle is the inherent heterogeneity of plastic particles, differences in polymer type, size, shape, surface chemistry, degree of environmental weathering, and toxicity can all influence particle behavior and interactions with co-contaminants. This variability complicates standardized sampling, quantification, and risk assessment efforts across environmental and biological systems. Several factors make MPs/NPs particularly difficult to study: (1) the variable degradation kinetics of different polymer types in natural environments, (2) the rate of fragmentation from macroplastics to MPs and NPs under environmental stressors, and (3) the duration and variability of exposure for different organisms. These dynamics are further obscured by limitations in current sampling methodologies, which often lack the sensitivity to detect the smallest NPs, particularly in complex matrices such as wastewater, seawater, and biological tissues.

Standardized criteria for defining and measuring particle size, shape, polymer identity, and surface functionality are still lacking. As a result, comparability between studies remains limited. While several analytical techniques, such as liquid chromatography–mass spectrometry, gas chromatography–mass spectrometry, flow cytometry, Fourier-transform infrared spectroscopy (FTIR), and various microscopy-based methods, have been employed for MP/NP detection [68, 115, 116], each has limitations in terms of resolution, throughput, or matrix interference. Environmental conditions during sampling can also influence the accuracy of results. For instance, studies that do not apply correction models for wind or wave action may underestimate or overestimate MP/NP concentrations. In some cases, plastic concentrations have been reported to be 2.5 times higher under windy conditions, underscoring the influence of local dynamics on surface accumulation [117].

6.5 Recommendations and policymakers

As the ecological and health impacts of MPs/NPs become increasingly evident, investment in MP/NP-specific treatment technologies has become as critical as their detection and monitoring in aquatic environments. NPs, in particular, often escape both conventional and advanced treatment systems due to their small size, necessitating the development of highly efficient and targeted removal technologies for both drinking water and wastewater systems. Research should prioritize the development of selective membrane technologies with tailored binding sites capable of adsorbing MPs/NPs with high specificity. Recent studies have demonstrated the potential of membrane-integrated processes to enhance NP removal from water and wastewater streams [118]. Parallel laboratory efforts should focus on designing bio-inspired flocculants, derived from environmentally benign and cost-effective materials, that aggregate MPs/NPs into larger flocs suitable for sedimentation or filtration. Advanced oxidation techniques, such as electrochemical oxidation, also show promise for breaking down plastic particles into non-toxic end products through the generation of reactive species like hydroxyl radicals. Furthermore, the implementation of multi-barrier hybrid systems, which integrate physical, chemical, and biological treatment processes, offers a robust approach for both contaminant degradation and MP/NP removals [111].

To protect aquatic ecosystems, remediation strategies should also target the *in situ* cleanup of plastic aggregates, including MPs/NPs. Innovations such as autonomous robotic systems and sensor-integrated detection technologies could be deployed to monitor and remove plastics directly from water bodies, especially in pollution hotspots. From a policy and sustainability standpoint, urgent action is needed to promote the production and adoption of biodegradable plastics that minimize the release of toxic additives. Regulatory measures should aim to restrict or phase out hazardous chemical additives known to contribute to plastic toxicity. In parallel, there must be greater emphasis on plastic recycling infrastructure, particularly for industrial applications, to prevent mismanagement leading to environmental release or open burning, which contributes to both air pollution and soil contamination. Overall, addressing the MPs/NPs crisis demands coordinated action among scientists, engineers, industry stakeholders, and policymakers, supported by international collaboration, public awareness, and evidence-based regulatory frameworks.

7. Conclusion

Extensive research efforts are currently focused on understanding the environmental behavior, health impacts, and remediation strategies related to MPs/NPs. This chapter provides a comprehensive overview of the toxicological risks posed by MPs/NPs to both human health and aquatic ecosystems, with a particular emphasis on their interactions with co-contaminants, including heavy metals, antibiotics, and pathogenic microorganisms. These plastic particles are increasingly found in food, drinking water, and atmospheric samples, thereby representing multiple exposure pathways for humans and wildlife. While conventional and advanced drinking water and wastewater treatment technologies, such as membrane filtration and media filtration, can achieve removal efficiencies exceeding 90%, residual MPs/NPs often remain in treated effluent. Challenges such as membrane fouling, pore saturation, and high operational costs limit the widespread implementation and long-term effectiveness of these systems. Furthermore, the persistence of MPs/NPs in treatment facilities and aquatic environments may enhance the mobility and toxicity of co-existing pollutants, amplifying ecological and health risks. Addressing MP/NP pollution requires holistic solutions that combine technological innovation, sustainable material design, and policy interventions. Strengthening regulatory frameworks, promoting international cooperation, and enforcing producer responsibility are essential steps toward minimizing plastic pollution and its far-reaching impacts. Continued interdisciplinary research will be crucial for advancing detection methods, improving treatment efficacy, and safeguarding public and environmental health from the threats posed by MPs and NPs.

Conflict of Interest

The authors declare no conflict of interest.

Author details

Kassim Chabi[1,2,3], Chengsong Ye[1,2], Mahmoud Gad[4], Neveen M. Rizk[4] and Xin Yu[1,2*]

1 College of the Environment & Ecology, Xiamen University, Xiamen, China

2 Fujian Key Laboratory of Coastal Pollution Prevention and Control, Xiamen University, Xiamen, China

3 Faculty of Sciences and Technic Abomey–Calavi, University of Abomey-Calavi, Cotonou, Benin

4 Water Pollution Research Department, National Research Centre, Giza, Egypt

*Address all correspondence to: xyu@xmu.edu.cn

References

[1] Sangkham S, Faikhaw O, Munkong N, Sakunkoo P, Arunlertaree C, Chavali M, et al. A review on microplastics and nanoplastics in the environment: Their occurrence, exposure routes, toxic studies, and potential effects on human health. Marine Pollution Bulletin. 2022;**181**:113832. DOI: 10.1016/j.marpolbul.2022.113832

[2] Arif Y, Mir AR, Zieliński P, Hayat S, Bajguz A. Microplastics and nanoplastics: Source, behavior, remediation, and multi-level environmental impact. Journal of Environmental Management. 2024;**356**:120618. DOI: 10.1016/j.jenvman.2024.120618

[3] Chen M, Khan AR, Memon MS, Iqbal B. Microplastics and nanoplastics across the food web: Challenges and mitigation strategies in securing human health. Process Safety and Environmental Protection. 2025;**201**:107586. DOI: 10.1016/j.psep.2025.107586

[4] Mahapatra S, Maity JP, Singha S, Mishra T, Dey G, Chandra Samal A, et al. Microplastics and nanoplastics in environment: Sampling, characterization and analytical methods. Groundwater for Sustainable Development. 2024;**26**:101267. DOI: 10.1016/j.gsd.2024.101267

[5] Yu Y, Mo WY, Luukkonen T. Adsorption behaviour and interaction of organic micropollutants with nano and microplastics – A review. Science of the Total Environment. 2021;**797**. 149140. DOI: 10.1016/j.scitotenv.2021.149140

[6] Rafa N, Ahmed B, Zohora F, Bakya J, Ahmed S, Ahmed SF, et al. Microplastics as carriers of toxic pollutants: Source, transport, and toxicological effects. Environmental Pollution. 2024;**343**:123190. DOI: 10.1016/j.envpol.2023.123190

[7] Du J, Huang W, Pan Y, Xu S, Li H, Jin M, et al. Ecotoxicological effects of microplastics combined with antibiotics in the aquatic environment: Recent developments and prospects. Environmental Toxicology and Chemistry Oxford University Press. 2024;**43**:1950–1961. DOI: 10.1002/etc.5950

[8] Bakan B, Kalčec N, Liu S, Ilić K, Qi Y, Capjak I, et al. Science-based evidence on pathways and effects of human exposure to micro- and nanoplastics. Archives of Industrial Hygiene and Toxicology, Walter de Gruyter GmbH. 2024;**75**:1–14

[9] Ogunola OS, Onada OA, Falaye AE. Mitigation measures to avert the impacts of plastics and microplastics in the marine environment (a review). Environmental Science and Pollution Research. 2018;**25**:9293–9310. DOI: 10.1007/s11356-018-1499-z

[10] Andrady AL, Neal MA. Applications and societal benefits of plastics. Philosophical Transactions of the Royal Society B: Biological Sciences. 2009;**364**(1526):1977–1984. DOI: 10.1098/rstb.2008.0304

[11] Hahladakis JN, Velis CA, Weber R, Iacovidou E, Purnell P. An overview of chemical additives present in plastics: Migration, release, fate and environmental impact during their use, disposal and recycling. Journal of

Hazardous Materials. 2018;**344**:179–199. DOI: 10.1016/j.jhazmat.2017.10.014

[12] Jambeck JR, Geyer R, Wilcox C, Siegler TR, Perryman M, Andrady A, et al. Plastic waste inputs from land into the ocean. Science. 2015;**347**:768–771

[13] Ali SS, Elsamahy T, Al-Tohamy R, Zhu D, Mahmoud YAG, Koutra E, . . . Sun J. Plastic wastes biodegradation: Mechanisms, challenges and future prospects. Science of the Total Environment. 2021;**780**:146590

[14] Zhang K, Hamidian AH, Tubić A, Zhang Y, Fang JKH, Wu C, et al. Understanding plastic degradation and microplastic formation in the environment: A review. Environmental Pollution. 2021;**274**:116554. DOI: 10.1016/j.envpol.2021.116554

[15] Oliveira YM, Vernin NS, Maia Bila D, Marques M, Tavares FW. Pollution caused by nanoplastics: Adverse effects and mechanisms of interaction via molecular simulation. Peer Journal. 2022;**10**:e13618. DOI: 10.7717/peerj.13618

[16] Lu T, Solis-Ramos E, Yi Y, Kumosa M. UV degradation model for polymers and polymer matrix composites. Polymer Degradation and Stability. 2018;**154**:203–210. DOI: 10.1016/j.polymdegradstab.2018.06.004

[17] Yousif E, Haddad R. Photodegradation and photostabilization of polymers, especially polystyrene: Review. SpringerPlus. 2013;**2**:398. DOI: 10.1186/2193-1801-2-398

[18] Sivasankar V, Sunitha TG, editors. Microplastics and Pollutants: Interactions, Degradations and Mechanisms. Nature Switzerland, Cham: Springer; 2024. DOI: 10.1007/978-3-031-54565-8

[19] Yan Y, Yu Y, Sima J, Geng C, Yang J. Aging behavior of microplastics accelerated by mechanical fragmentation: Alteration of intrinsic and extrinsic properties. Environmental Science and Pollution Research. 2023;**30**:90993–91006. DOI: 10.1007/s11356-023-28736-x

[20] Dintcheva NT. Overview of polymers and biopolymers degradation and stabilization towards sustainability and materials circularity. Polymer. 2024;**306**:127136. DOI: 10.1016/j.polymer.2024.127136

[21] Eddy, N.A comprehensive review on polymer degradation: mechanisms, environmental implications, and sustainable mitigation strategies. Communication in Physical Sciences. 2025;**12**(3):799–807

[22] Chen -C-C, Dai L, Ma L, Guo R-T. Enzymatic degradation of plant biomass and synthetic polymers. Nature Reviews Chemistry. 2020;**4**:114–126. DOI: 10.1038/s41570-020-0163-6

[23] Suresh V, Shams R, Dash KK, Shaikh AM, Béla K. Comprehensive review on enzymatic polymer degradation: A sustainable solution for plastics. Journal of Agriculture and Food Research. 2025;**20**: 101788. DOI: 10.1016/j.jafr.2025.101788

[24] Wiesinger H, Wang Z, Hellweg S. Deep dive into plastic monomers, additives, and processing aids. Environmental Science & Technology. 2021;**55**:9339–9351. DOI: 10.1021/acs.est.1c00976

[25] Płuciennik K, Sicińska P, Misztal W, Bukowska B. Important factors affecting

induction of cell death, oxidative stress and DNA Damage by nano- and microplastic particles in vitro. Cells. 2024;**13**:768. DOI: 10.3390/cells13090768

[26] Mattioda V, Benedetti V, Tessarolo C, Oberto F, Favole A, Gallo M, et al. Pro-Inflammatory and cytotoxic effects of polystyrene microplastics on human and murine intestinal cell lines. Biomolecules. 2023;**13**:140

[27] Grande-Tovar CD, Carrascal-Hernández DC, Trilleras J, Mora K, Arana VA. Microplastics' and nanoplastics' interactions with microorganisms: A bibliometric study. Sustainability. 2022;**14**:14761. DOI: 10.3390/su142214761

[28] Wang R, Li X, Li J, Dai W, Luan Y. Bacterial Interactions with nanoplastics and the environmental effects they cause. Fermentation. 2023;**9**:939. DOI: 10.3390/fermentation9110939

[29] Shen M, Zhu Y, Zhang Y, Zeng G, Wen X, Yi H, et al. Micro(nano)plastics: Unignorable vectors for organisms. Marine Pollution Bulletin. 2019;**139**:328–331

[30] Gao W, Xu M, Zhao W, Yang X, Xin F, Dong W, et al. Microbial degradation of (Micro)plastics: Mechanisms, enhancements, and future directions. Fermentation. 2024;**10**:441

[31] Ghosh S, Qureshi A, Purohit HJ. Microbial degradation of plastics: Biofilms and degradation pathways. In: Contaminants in Agriculture and Environment: Health Risks and Remediation. Haridwar, India: Agro Environ Media - Agriculture and Ennvironmental Science Academy; 2019. p. 184–199. DOI: 10.26832/aesa-2019-cae-0153-014

[32] Stapleton MJ, Ansari AJ, Hai FI. Antibiotic sorption onto microplastics in water: A critical review of the factors, mechanisms and implications. Water Research. 2023;**233**:119790. DOI: 10.1016/j.watres.2023.119790

[33] Zhuang S, Wang J. Interaction between antibiotics and microplastics: Recent advances and perspective. Science of the Total Environment. 2023;**897**:165414. DOI: 10.1016/j.scitotenv.2023.165414

[34] Liu Z, Liao L, Yang H, Chen Z, Ji H. Impact of micro-nano plastics on microorganisms and thus drinking water treatment effectiveness: A review. Desalination and Water Treatment. 2024;**320**:100673. DOI: 10.1016/j.dwt.2024.100673

[35] Liu S, Ding H, Song Y, Xue Y, Bi M, Wu M, et al. The potential risks posed by micro-nanoplastics to the safety of disinfected drinking water. Journal of Hazardous Materials. 2023;**450**:131089. DOI: 10.1016/j.jhazmat.2023.131089

[36] Zhang X, Liu L, Chen X, Li J, Chen J, Wang H. The fate and risk of microplastic and antibiotic sulfamethoxazole coexisting in the environment. Environmental Geochemistry and Health. 2023;**45**:2905–2915. DOI: 10.1007/s10653-022-01385-8

[37] Chen P-W, Hsiao M-N, Xiao L-W, Liu Z-S. Adsorption behavior of heavy metals onto microplastics derived from conventional and biodegradable commercial plastic products. Science of the Total Environment. 2024;**951**:175537. DOI: 10.1016/j.scitotenv.2024.175537

[38] Wang Q, Zhang Y, Wangjin X, Wang Y, Meng G, Chen Y. The

adsorption behavior of metals in aqueous solution by microplastics effected by UV radiation. Journal of Environmental Sciences. 2020;**87**: 272–280. DOI: 10.1016/j.jes.2019.07.006

[39] Lin L, Li H, Hong H, Yuan B, Sun X, He L, et al. Enhanced heavy metal adsorption on microplastics by incorporating flame retardant hexabromocyclododecanes: Mechanisms and potential migration risks. Water Research. 2022;**225**:119144. DOI: 10.1016/j.watres.2022.119144

[40] Wang Y, Wang Y, Liu M, Jia R, Zhang Y, Sun G, et al. Micro-/nano-plastics as vectors of heavy metals and stress response of ciliates using transcriptomic and metabolomic analyses. Environmental Pollution. 2024;**360**:124667. DOI: 10.1016/j.envpol.2024.124667

[41] Wang T, Wang L, Chen Q, Kalogerakis N, Ji R, Ma Y. Interactions between microplastics and organic pollutants: Effects on toxicity, bioaccumulation, degradation, and transport. Science of the Total Environment. 2020;**748**:142427. DOI: 10.1016/j.scitotenv.2020.142427

[42] Bhagat J, Nishimura N, Shimada Y. Toxicological interactions of microplastics/nanoplastics and environmental contaminants: Current knowledge and future perspectives. Journal of Hazardous Materials. 2021;**405**. 123913. DOI: 10.1016/j.jhazmat.2020.123913

[43] Oßmann BE, Sarau G, Holtmannspötter H, Pischetsrieder M, Christiansen SH, Dicke W. Small-sized microplastics and pigmented particles in bottled mineral water. Water Research. 2018;**141**:307–316. DOI: 10.1016/j.watres.2018.05.027

[44] Barbara.E.O.ßmann. Microplastics in drinking water ? Present state of knowledge and open questions. Current Opinion in Food Science. 2021;**41**:45–51. DOI: 10.1016/j.cofs.2021.02.011

[45] Maharjan KK. Microplastic pollution in bottled water: A systematic review. International Journal of Environmental Science and Technology. 2025;**22**:1283–1296. DOI: 10.1007/s13762-024-05807-1

[46] Gao W, Zhang Y, Mo A, Jiang J, Liang Y, Cao X. Removal of microplastics in water: Technology progress and green strategies. Green Analytical Chemistry. 2022;**3**:100042–100042. DOI: 10.1016/j.greeac.2022.100042

[47] Strady E, Kieu-Le T-C, Gasperi J, Tassin B. Temporal dynamic of anthropogenic fibers in a tropical river-estuarine system. Environmental Pollution. 2020;**259**. 113897. DOI: 10.1016/j.envpol.2019.113897

[48] Wang R, Tan H, Huang X, Wang J, Su W, Maqbool F, et al. Impact of micro- and nano-plastics on marine organisms under environmentally relevant conditions. Aquatic Toxicology. 2025;**286**:107475. DOI: 10.1016/j.aquatox.2025.107475

[49] Xu Y, Ou Q, Wang X, Hou F, Li P, Van Der Hoek JP, et al. Assessing the mass concentration of microplastics and nanoplastics in wastewater treatment plants by pyrolysis gas chromatography–mass spectrometry. Environmental Science & Technology. 2023;**57**:3114–3123

[50] Jeong E, Lee J-Y, Redwan M. Animal exposure to microplastics and health effects: A review. Emerging

Contaminants. 2024;**10**. 100369. DOI: 10.1016/j.emcon.2024.100369

[51] Bui X-T, Vo TDH, Nguyen P-T, Nguyen V-T, Dao T-S, Nguyen P-D. Microplastics pollution in wastewater: Characteristics, occurrence and removal technologies. Environmental Technology & Innovation. 2020;**19**. 101013. DOI: 10.1016/j.eti.2020.101013

[52] Praveena SM, Shamsul Ariffin NI, Nafisyah AL. Microplastics in Malaysian bottled water brands: Occurrence and potential human exposure. Environmental Pollution. 2022;**315**. 120494. DOI: 10.1016/j.envpol.2022.120494

[53] Zhou X, Wang J, Li H, Zhang H, Hua-Jiang, Zhang DL. Microplastic pollution of bottled water in China. Journal of Water Process Engineering. 2021;**40**. 101884. DOI: 10.1016/j.jwpe.2020.101884

[54] Marfella R, Prattichizzo F, Sardu C, Fulgenzi G, Graciotti L, Spadoni T, et al. Microplastics and Nanoplastics in Atheromas and Cardiovascular Events. New England Journal of Medicine. 2024;**390**:900–910

[55] Leslie HA, Van Velzen MJM, Brandsma SH, Vethaak AD, Garcia-Vallejo JJ, Lamoree MH. Discovery and quantification of plastic particle pollution in human blood. Environment International. 2022;**163**. 107199. DOI: 10.1016/j.envint.2022.107199

[56] Nihart AJ, Garcia MA, El Hayek E, Liu R, Olewine M, Kingston JD, et al. Bioaccumulation of microplastics in decedent human brains. Nature Medicine. 2025;**31**:1114–1119

[57] Boran T, Zengin OS, Seker Z, Akyildiz AG, Kara M, Oztas E, et al. An evaluation of a hepatotoxicity risk induced by the microplastic polymethyl methacrylate (PMMA) using HepG2/THP-1 co-culture model. Environmental Science and Pollution Research. 2024;**31**:28890–28904

[58] Schmid O, Cassee FR. On the pivotal role of dose for particle toxicology and risk assessment: Exposure is a poor surrogate for delivered dose. Particle and Fibre Toxicology. 2017;**14**(52):s12989–017–0233–1. DOI: 10.1186/s12989-017-0233-1

[59] Danopoulos E, Twiddy M, Rotchell JM. Microplastic contamination of drinking water: A systematic review. PLOS ONE. 2020;**15**(7):1–23. DOI: 10.1371/journal.pone.0236838

[60] Mccormick A, Hoellein TJ, Mason SA, Schluep J, Kelly JJ. Microplastic is an abundant and distinct microbial habitat in an Urban River. Environmental Science & Technology. 2014;**48**:11863–11871–11863–11871. DOI: 10.1021/es503610r

[61] Kirstein IV, Gomiero A, Vollertsen J. Microplastic pollution in drinking water. Current Opinion in Toxicology. 2021;**28**:70–75. DOI: 10.1016/j.cotox.2021.09.003

[62] Pulido-reyes G, Magherini L, Bianco C, Sethi R, Von U, Kaegi R, et al. Nanoplastics removal during drinking water treatment: Laboratory- and pilot-scale experiments and modeling. Journal of Hazardous Materials. 2022;**436**:129011–129011

[63] De Voogt P, editor. Reviews of Environmental Contamination and Toxicology. Vol. **259** Cham: Springer International Publishing, 2021. DOI: 10.1007/978-3-030-88342-3

[64] Habumugisha T, Zhang Z, Uwizewe C, Yan C, Ndayishimiye JC, Rehman A, et al. Toxicological review of micro- and nano-plastics in aquatic environments: Risks to ecosystems, food web dynamics and human health. Ecotoxicology and Environmental Safety. 2024;**278**:116426. DOI: 10.1016/j.ecoenv.2024.116426

[65] Thakur R, Joshi V, Sahoo GC, Jindal N, Tiwari RR, Rana S. Review of mechanisms and impacts of nanoplastic toxicity in aquatic organisms and potential impacts on human health. Toxicology Reports. 2025;**14**:102013. DOI: 10.1016/j.toxrep.2025.102013

[66] Ziani K, Ioniță-Mîndrican C-B, Mititelu M, Neacșu SM, Negrei C, Moroșan E, et al. Microplastics: A real global threat for environment and food safety: A state of the art review. Nutrients. 2023;**15**:617

[67] Zhang N, Li YB, He HR, Zhang JF, Ma GS. You are what you eat: Microplastics in the feces of young men living in Beijing. Science of the Total Environment. 2021;**767**. 144345. DOI: 10.1016/j.scitotenv.2020.144345

[68] Ragusa A, Svelato A, Santacroce C, Catalano P, Notarstefano V, Carnevali O, ... Giorgini E. Plasticenta: First evidence of microplastics in human placenta. Environment International. 2021;**146**:106274

[69] Li Y, Chen L, Zhou N, Chen Y, Ling Z, Xiang P. Microplastics in the human body: A comprehensive review of exposure, distribution, migration mechanisms, and toxicity. Science of the Total Environment. 2024;**946**:174215

[70] Saha U, Kumari P, Ghosh A, Sinha A, Jena S, Kirti A, . . . Verma SK. Detrimental consequences of micropolymers associated plasticizers on endocrinal disruption. Materials Today Bio. 2024;**27**:101139

[71] Alqahtani S, Alqahtani S, Saquib Q, Mohiddin F. Toxicological impact of microplastics and nanoplastics on humans: Understanding the mechanistic aspect of the interaction. Frontiers in Toxicology. 2023;**5**. DOI: 10.3389/ftox.2023.1193386

[72] Maddela NR, Kakarla D, Venkateswarlu K, Megharaj M. Additives of plastics: Entry into the environment and potential risks to human and ecological health. Journal of Environmental Management. 2023;**348**. 119364. DOI: 10.1016/j.jenvman.2023.119364

[73] Shekh MR, Kumar V. Impact of plastic pollution on ecosystems: A review of adverse effects and sustainable solutions. In: Environ Monit Assess. : Environmental Monitoring and Assessment; 2025. p. 197. DOI: 10.1007/s10661-025-13723-1

[74] Saavedra J, Stoll S, Slaveykova VI. Influence of nanoplastic surface charge on eco-Corona formation, aggregation and toxicity to freshwater zooplankton. Environmental Pollution. 2019;**252**: 715–722. DOI: 10.1016/j.envpol.2019.05.135

[75] Geremia E, Muscari Tomajoli MT, Murano C, Petito A, Fasciolo G. The impact of micro- and nanoplastics on aquatic organisms: mechanisms of oxidative stress and implications for human health—a review. Environments. 2023;**10**:161. DOI: 10.3390/environments10090161

[76] Aransiola SA, Victor-Ekwebelem MO, Daza BX, Oladoye PO, Alli YA, Bamisaye A, et al. Micro- and

nano-plastics pollution in the marine environment: Progresses, drawbacks and future guidelines. Chemosphere. 2025;**374**:144211. DOI: 10.1016/j.chemosphere.2025.144211

[77] Al-Thawadi S. Microplastics and nanoplastics in aquatic environments: challenges and threats to aquatic organisms. Arabian Journal for Science and Engineering. 2020;**45**:4419–4440. DOI: 10.1007/s13369-020-04402-z

[78] Huang H, Hou J, Li M, Wei F, Liao Y, Xi B. Microplastics in the bloodstream can induce cerebral thrombosis by causing cell obstruction and lead to neurobehavioral abnormalities. Science Advances. 2025;**11**. DOI: 10.1126/sciadv.adr8243

[79] SAPEA (Science Advice for Policy by European Academies). A Scientific Perspective on Microplastics in Nature and Society: Evidence Review Report. Berlin: SAPEA; 2019. DOI: 10.26356/microplastics

[80] World Health Organization. Microplastics in Drinking-water. Geneva: WHO; 2019, ISBN: 978-92-4-151619-8. https://iris.who.int/handle/10665/326499

[81] Koelmans AA, Mohamed Nor NH, Hermsen E, Kooi M, Mintenig SM, De France J. Microplastics in freshwaters and drinking water: Critical review and assessment of data quality. Water Research. 2019;**155**:410–422. DOI: 10.1016/j.watres.2019.02.054

[82] De Ruijter VN, Redondo-Hasselerharm PE, Gouin T, Koelmans AA. Quality criteria for microplastic effect studies in the context of risk assessment: a critical review. Environmental Science & Technology. 2020;**54**:11692–11705. DOI: 10.1021/acs.est.0c03057

[83] Moyo S. An enigma: A meta-analysis reveals the effect of ubiquitous microplastics on different taxa in aquatic systems. Frontiers in Environmental Science. 2022;**10**. 999349. DOI: 10.3389/fenvs.2022.999349

[84] Foley CJ, Feiner ZS, Malinich TD, Höök TO. A meta-analysis of the effects of exposure to microplastics on fish and aquatic invertebrates. Science of the Total Environment. 2018; **631-632**: 550–559. DOI: 10.1016/j.scitotenv.2018.03.046

[85] Zantis LJ, Carroll EL, Nelms SE, Bosker T. Marine mammals and microplastics: A systematic review and call for standardisation. Environmental Pollution. 2021;**269**. 116142. DOI: 10.1016/j.envpol.2020.116142

[86] Subair A, G M, Chellappan S, Chinglenthoiba C, S IM. Techniques for removal and degradation of microplastics. In: Microplastic Occurrence, Fate, Impact, and Remediation. Cham: Springer Nature Switzerland; 2023. p. 127–153

[87] Pivokonsky M, Cermakova L, Novotna K, Peer P, Cajthaml T, Janda V. Occurrence of microplastics in raw and treated drinking water. Science of the Total Environment. 2018;**643**: 1644–1651

[88] Charlton-Howard HS, Bond AL, Rivers-Auty J, Lavers JL. 'Plasticosis': Characterising macro- and microplastic-associated fibrosis in seabird tissues. Journal of Hazardous Materials. 2023;**450**:131090. DOI: 10.1016/j.jhazmat.2023.131090

[89] Chabi K, Li J, Ye C, Kiki C, Xiao X, Li X, et al. Rapid sand filtration for <10 μm-sized microplastic removal in tap

water treatment: Efficiency and adsorption mechanisms. Science of the Total Environment. 2024;**912**. DOI: 10.1016/j.scitotenv.2023.169074

[90] Wang Y, Yang Y, Liu X, Zhao J, Liu R, Xing B. Interaction of microplastics with antibiotics in aquatic environment: Distribution, adsorption, and toxicity. Environmental Science & Technology. 2021;**55**:15579–15595. DOI: 10.1021/acs.est.1c04509

[91] Liu Q, Khor SM. Emerging absorption-based techniques for removing microplastics and nanoplastics from actual water bodies. TrAC Trends in Analytical Chemistry. 2024;**170**. 117465. DOI: 10.1016/j.trac.2023.117465

[92] Keerthana Devi M, Karmegam N, Manikandan S, Subbaiya R, Song H, Kwon EE, et al. Removal of nanoplastics in water treatment processes: A review. Science of the Total Environment. 2022;**845**:157168. DOI: 10.1016/j.scitotenv.2022.157168

[93] Hofman-Caris CHM, Bäuerlein PS, Siegers WG, Mintenig SM, Messina R, Dekker SC, et al. Removal of nanoparticles (both inorganic nanoparticles and nanoplastics) in drinking water treatment – Coagulation/flocculation/sedimentation, and sand/granular activated carbon filtration. Environmental Science: Water Research & Technology Royal Society of Chemistry. 2022. DOI: 10.1039/d2ew00226d

[94] Ganie ZA, Khandelwal N, Tiwari E, Singh N, Darbha GK. Biochar-facilitated remediation of nanoplastic contaminated water: Effect of pyrolysis temperature induced surface modifications. Journal of Hazardous Materials. 2021;**417**. 126096. DOI: 10.1016/j.jhazmat.2021.126096

[95] Ali I, Tan X, Mustafa G, Gao J, Peng C, Naz I, et al. Removal of micro- and nanoplastics by filtration technology: Performance and obstructions to market penetrations. Journal of Cleaner Production. 2024;**470**:143305. DOI: 10.1016/j.jclepro.2024.143305

[96] Li Q, Lai Y, Yu S, Li P, Zhou X, Dong L, et al. Sequential Isolation of microplastics and nanoplastics in environmental waters by membrane filtration, followed by cloud-point extraction. Analytical Chemistry. 2021;**93**:4559–4566

[97] Subair A, Priya KL, Chellappan S, Rahuman AT, Hridya J, Devi PS, et al. Evaluating the performance of electrocoagulation system in the removal of polystyrene microplastics from water. Environmental Research. 2024;**243**:117887. DOI: 10.1016/j.envres.2023.117887

[98] Modak S, Kasula M, Esfahani MR. Nanoplastics Removal from Water using Metal–Organic Framework: Investigation of Adsorption Mechanisms, Kinetics, and Effective Environmental Parameters. ACS Applied Engineering Materials. 2023;**1**:744–755. DOI: 10.1021/acsaenm.2c00174

[99] Laishram D, Kim S, Lee S, Park S. Advancements in biochar as a sustainable adsorbent for water pollution mitigation. Advanced Science. 2025;**12**. 2410383. DOI: 10.1002/advs.202410383

[100] Mulindwa P, Kasule JS, Nantaba F, Wasswa J, Expósito AJ. Bioadsorbents for removal of microplastics from water

ecosystems: A review. International Journal of Sustainable Engineering. 2024;**17**:582–599. DOI: 10.1080/19397038.2024.2374003

[101] Zheng H, Chen Q, Chen Z. Carbon-based adsorbents for micro/nano-plastics removal: Current advances and perspectives. Water Emerging Contaminants & Nanoplastics. 2024;**3**. DOI: 10.20517/wecn.2023.74

[102] Tang W, Li H, Fei L, Wei B, Zhou T, Zhang H. The removal of microplastics from water by coagulation: A comprehensive review. Science of the Total Environment. 2022;**851**. 158224. DOI: 10.1016/j.scitotenv.2022.158224

[103] Bodzek M, Bodzek P. Remediation of micro- and nanoplastics by membrane technologies. Membranes. 2025;**15**:82. DOI: 10.3390/membranes15030082

[104] Al-Asheh S, Bagheri M, Aidan A. Membrane bioreactor for wastewater treatment: A review. Case Studies in Chemical and Environmental Engineering. 2021;**4**. 100109. DOI: 10.1016/j.cscee.2021.100109

[105] Poerio T, Piacentini E, Mazzei R. Membrane processes for microplastic removal. Molecules. 2019;**24**:4148. DOI: 10.3390/molecules24224148

[106] Khan MT, Ahmad M, Hossain MF, Nawab A, Ahmad I, Ahmad K, et al. Microplastic removal by coagulation: A review of optimizing the reaction conditions and mechanisms. Water Emerging Contaminants & Nanoplastics. 2023:2. DOI: 10.20517/wecn.2023.39

[107] Fazli A, Athanassiou A, Fragouli D. Current approaches and challenges in advanced oxidation processes for nanoplastic degradation. Chemistry. 2025. DOI: 10.26434/chemrxiv-2025-r15f3

[108] Chen R, Zhang Z, Deng Y, Wang J, Cui Y, Zhao Y, et al. Electrochemical degradation of PET microplastics and its mechanism. Industrial & Engineering Chemistry Research. 2025;**64**:7915–7929

[109] Kiendrebeogo M, Karimi Estahbanati MR, Ouarda Y, Drogui P, Tyagi RD. Electrochemical degradation of nanoplastics in water: Analysis of the role of reactive oxygen species. Science of the Total Environment. 2022;**808**. 151897. DOI: 10.1016/j.scitotenv.2021.151897

[110] Akash K, Parthasarathi R, Kanmani Bharathi J, Elango R. Exploring the Role of Microbes in the Biodegradation of Plastic Waste: Mechanisms, Interactions, and Implications for Sustainable Waste Management-A Review. Water, Air, and Soil Pollution. 2025;**236**:786. DOI: 10.1007/s11270-025-08343-x

[111] Romphophak P, Faikhaw O, Sairiam S, Thuptimdang P, Coufort-Saudejaud C. Removal of microplastics and nanoplastics in water treatment processes: A systematic literature review. Journal of Water Process Engineering. 2024;**64**. 105669. DOI: 10.1016/j.jwpe.2024.105669

[112] Omura T, Isobe N, Miura T, Ishii S, Mori M, Ishitani Y, et al. Microbial decomposition of biodegradable plastics on the deep-sea floor. Nature Communications. 2024;**15**:568

[113] Sahith VN, Kumar A, Sruthi VS, Sathish S, Venkatesan D, Prabu D, Samrot A. Microbial and Enzymatic biodegradation of microplastics and

nanoplastics: Advances, challenges, and sustainable solutions for environmental remediation. Desalination and Water Treatment. 2025;101450

[114] Rasheed A, Bashir Y, Das S. Are (bio)electrochemical techniques sustainable solutions to combat micro- and nano-plastic pollution? Current Opinion in Chemical Engineering. 2025;**47**. 101091. DOI: 10.1016/j.coche.2025.101091

[115] Qiu Q, Tan Z, Wang J, Peng J, Li M, Zhan Z. Extraction, enumeration and identification methods for monitoring microplastics in the environment. Estuarine, Coastal and Shelf Science. 2016;**176**:102–109

[116] Singh S, Kalyanasundaram M, Diwan V. Removal of microplastics from wastewater: Available techniques and way forward. Water Science and Technology. 2021;**84**:3689–3704. DOI: 10.2166/wst.2021.472

[117] Boucher J, Billard G. The challenges of measuring plastic pollution. Field Actions Science Reports. The Journal of Field Actions. 2019;(Special Issue 19):68–75

[118] Sutrisna PD, Riadi L, Buana PCW, Khoiruddin K, Boopathy R, Wenten IG, et al. Membrane and membrane-integrated processes for nanoplastics removal and remediation. Polymer Degradation and Stability. 2024;**220**:110635. DOI: 10.1016/j.polymdegradstab.2023.110635

Chapter 5

Perspective Chapter: Environmental Ethics and the Microplastic Crisis – A Philosophical Inquiry into Human Responsibility and Aquatic Justice

Egara Wilfred Oviemuno, Etaware Peter Mudiaga and Oviemuno Bridget Gwa

Abstract

This chapter examines the microplastic crisis through the dual lenses of environmental ethics and empirical science by emphasizing the intersections of human responsibility and aquatic justice. Microplastics, which are best described as plastic particles smaller than 5 mm, are now recognized as one of the most pervasive pollutants of the twenty-first century, with an estimated 15–51 trillion particles in global oceans and 8–10 million tonnes entering annually. Their persistence in aquatic systems leads to ingestion by over 100 marine species, reduced feeding efficiency in copepods by up to 40%, and measurable risks to human health via seafood consumption and tap water contamination. Through the hermeneutical method of enquiry and a critical analysis of anthropocentric, biocentric, ecocentric, and interspecies justice perspectives, this chapter interrogates humanity's moral responsibility toward aquatic ecosystems. It proposes the concept of aquatic justice, extending environmental justice principles to non-human life and future generations, and grounding ethical obligations in intergenerational equity, intrinsic ecosystem value, and restorative responsibility. Hence, by linking scientific evidence with ethical reasoning, the chapter calls for coordinated interventions such as policy reform, corporate accountability, technological innovation, and cultural reorientation that prioritize ecological integrity and justice. Ultimately, this work contributes to ongoing discussions on sustainability by situating the microplastic crisis not only as an environmental issue but also as a profound ethical challenge requiring collective and transformative action.

Keywords: environmental ethics, microplastic crisis, human responsibility, aquatic justice, sustainability, anthropocentrism

1. Introduction

Microplastics can be described as plastic particles that are smaller than 5 mm in diameter and are found in aquatic ecosystems, which represent one of the most pressing environmental challenges of the twenty-first century [1]. They emerge as a critical environmental pollutant in aquatic ecosystems, including oceans, rivers, and lakes [1]. These particles, originating from primary sources such as microbeads in cosmetics and secondary sources such as degraded plastic waste, are estimated to number between 15 and 51 trillion in global oceans [2]. Their persistence in marine and freshwater environments disrupts ecosystems by causing physical harm to organisms through ingestion and entanglement and by facilitating the transfer of toxic pollutants through food chains, posing risks to both biodiversity and human health [3, 4]. For instance, microplastics have been detected in over 100 marine species, with potential impacts on human health via seafood consumption [5]. Beyond ecological consequences, the microplastic crisis prompts ethical considerations regarding human responsibility and the moral obligations owed to aquatic environments [6]. This chapter, therefore, explores the crisis through the lens of environmental ethics, integrating frameworks such as anthropocentrism, biocentrism, and ecocentrism to propose a concept of aquatic justice that recognizes the intrinsic value of water-based ecosystems [7]. By combining empirical data with ethical theories, this work aims to articulate moral imperatives for mitigating microplastic pollution and fostering sustainable human-nature relationships, emphasizing environmental justice and intergenerational equity [8]. The discussion contributes to broader debates about humanity's role in addressing environmental limits, advocating for practices that prioritize ecological integrity and responsibility.

2. The microplastic crisis: Scope and impact

Microplastics are defined as plastic particles smaller than 5 mm in diameter and are seen as a pervasive pollutant in aquatic ecosystems, encompassing both primary microplastics (e.g., microbeads in cosmetics and synthetic fibers from textiles) and secondary microplastics (e.g., fragments from degraded larger plastics) [1]. These particles are ubiquitous, detected in diverse environments from Arctic ice to deep-sea sediments, with global ocean estimates ranging from 15 to 51 trillion particles [2, 9]. Primary microplastics enter aquatic systems directly through wastewater from consumer products, such as cosmetics and laundry effluents, while secondary microplastics result from the breakdown of mismanaged plastic waste via photo-degradation, mechanical abrasion, or weathering [10]. Major pathways into marine and freshwater environments include urban runoff, wastewater treatment plant effluents, atmospheric deposition, and direct littering, with rivers acting as significant conduits, transporting over 1.15 to 2.41 million tonnes of plastic to oceans annually [11]. Comparative studies highlight regional variations: Lebreton et al. [11] estimate Asia contributes 67% of riverine plastic inputs, while Jambeck et al. [12] report North America's contribution at 0.9–1.8% [11, 12].

Ecologically, microplastics' persistence (100–1,000 years) enables ingestion by over 100 marine species, from zooplankton to cetaceans, causing physical harm (e.g., 40% reduced feeding efficiency in copepods) and chemical contamination via toxin bioaccumulation [13, 14]. For instance, Wright et al. [14] found that microplastics increase mortality in copepods by 30%, while Gall and Thompson [15] reported up to 20% mortality in fish at high concentrations [14, 15]. Human health risks include inflammation and endocrine disruption, with microplastics detected in 90% of seafood and 83% of global tap water samples [16]. The crisis's scale, with 8–10 million tonnes of plastic entering oceans annually, underscores the need for systemic solutions [12]. The global scale of the crisis, coupled with its long-term environmental and health implications, underscores the need for improved waste management, regulatory measures, and interdisciplinary solutions to mitigate microplastic pollution [9].

More so, advances in microplastic research have enhanced detection accuracy. Fourier-transform infrared spectroscopy (FTIR) and Raman spectroscopy identify polymer types down to 1 μm with 95% accuracy, which surpasses visual sorting methods that underestimate particles below 500 μm [17]. Automated sampling, too, such as continuous-flow centrifugation, captures 80–90% of microplastics in water columns compared to 50–60% for traditional manta nets [18]. Comparative analysis also reveals methodological trade-offs: For example, Hidalgo-Ruz et al. [19] noted that spectroscopic methods are precise but costly, while, on the other hand, net sampling is scalable but less accurate for smaller particles [19]. Standardized protocols, as proposed by GESAMP (2021), are critical for data comparability [20]. Hence, while scientific advancements improve monitoring, the microplastic crisis demands ethical and systemic interventions, as explored in subsequent sections.

Table 1 identifies the primary human-related sources of microplastics and describes how they enter the environment, especially aquatic systems. The table highlights how everyday products and industrial activities serve as critical sources of microplastic pollution, thereby emphasizing the need for better waste management and regulatory controls.

3. Environmental ethics: Frameworks for understanding responsibility

Environmental ethics, therefore, provides three philosophical foundations for analyzing the microplastic crisis. These three primary ethical frameworks, which include anthropocentrism, biocentrism, and ecocentrism, offer distinct perspectives on human responsibility toward aquatic ecosystems. Below, we will look at them to understand the concepts and their roles in the aquatic space.

3.1 Anthropocentrism

Anthropocentrism prioritizes human interests, viewing nature as a resource for human use. From this perspective, the microplastic crisis is seen as problematic primarily because it threatens human health, food security, and economic interests, such as fisheries and tourism, as opined by Beaumont et al. [25]. Anthropocentric ethics might justify actions that mitigate microplastics based on self-interest, such as reducing health risks or preserving ecosystem services. However, this framework risks overlooking the intrinsic value of non-human life and ecosystems, which limits the scope of moral consideration to human welfare.

Source	Description	Entry pathway	Reference
Synthetic textiles (primary)	Synthetic fibers are released during washing.	Wastewater from laundry effluents.	[10]
Personal care products (primary)	Microbeads in toothpaste, scrubs, and cosmetics.	Household wastewater often bypasses treatment.	[1]
Plastic waste fragmentation (secondary)	Degraded plastic bags, bottles, and containers, through weathering or abrasion.	Land-based runoff, river transport, and direct dumping.	[11]
Industrial spillage	Resin pellets from plastic manufacturing.	Industrial discharge into waterways.	[21]
Vehicle tires and paints (secondary)	Abrasion of synthetic rubber tires and weathering of paints.	Urban runoff and road drainage into rivers and coastal waters.	[22]
Maritime activities	Paint flakes, cargo loss, and plastic debris from shipping or oil/gas operations.	Direct release into marine environments.	[20]
Fishing activities	Lost or abandoned fishing gear (e.g., nets, lines) fragments into microplastics.	Direct input from recreational and industrial fishing.	[23]
Aquaculture	Plastic equipment (e.g., cages, ropes) degrading into microplastics.	Direct release into coastal or freshwater systems.	[5]
Tourism/recreation	Single-use plastics (e.g., straws, cups) resulting from coastal tourism activities.	Littering and runoff into marine/coastal waters.	[24]

Table 1.
Major sources of microplastics and their environmental entry points.

3.2 Biocentrism

Biocentrism extends moral obligation and consideration to all living things or organisms, recognizing their inherent value regardless of their utility to humans [26]. In the context of microplastics, biocentrism emphasizes the harm inflicted on marine species, from zooplankton to cetaceans, which suffer physical and physiological damage from plastic ingestion [15]. This biocentric perspective demands that humans take responsibility for reducing microplastic pollution to protect the well-being of individual organisms. This perspective challenges the anthropocentric focus on human-centric outcomes. This will be seen in **Figure 1** chart below.

3.3 Ecocentrism

Different from the other two perspectives, Ecocentrism takes a holistic view, valuing entire ecosystems (comprising both humans and micro-organisms) and their interdependent relationships [27]. From an ecocentric perspective, the microplastic crisis disrupts the integrity and stability of aquatic ecosystems, compromising biodiversity and ecological processes. This perspective calls for systemic changes to address the root causes of microplastic pollution, such as the overproduction of

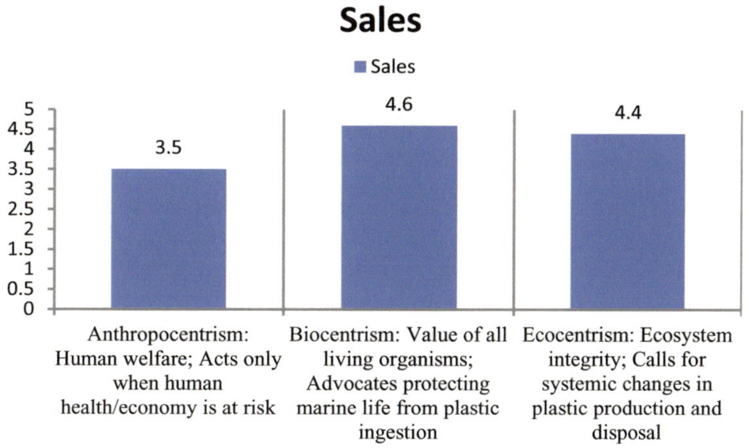

Figure 1.
Ethical frameworks in environmental decision-making.

plastics and inadequate waste management systems. Ecocentrism aligns with the concept of aquatic justice, which is explored later in this chapter.

From the foregoing, it is deduced that one must look at all dimensions and possibilities before taking decisions that have to do with the environment in a bid to protect it, that is, both land and aquatic ecosystems for justice's sake. Thus, these ethical frameworks are necessary as they point toward that which must be prioritized for the sake of justice, as shown in **Figure 1**.

4. Human responsibility and the microplastic crisis

The microplastic crisis is a direct consequence of human actions, resulting from industrial production and consumer behavior. Philosophically, this raises questions about moral responsibility and accountability. Immanuel Kant's categorical imperative, which urges individuals to act according to principles that could be universalized [28], suggests that the unchecked production and disposal of plastics violate ethical duties to avoid harm at all levels. Similarly, Hans Jonas' principle of responsibility emphasizes the obligation to protect future generations and the environment in light of humanity's technological power [29]. The pervasive nature of microplastics, which transcend national boundaries and persist across generations, underscores the need for collective and forward-looking responsibility.

More so, the concept of "slow violence," according to Nixon, is particularly relevant [30]. Microplastic pollution represents a form of incremental, often invisible harm that accumulates over time, disproportionately affecting vulnerable ecosystems and communities. This temporal and spatial disconnect between cause and effect complicates the attribution of responsibility, yet it reinforces the ethical imperative to act pre-emptively to prevent further harm to living organisms and their environments.

5. Aquatic justice: A new ethical paradigm

To address the microplastic crisis comprehensively, this chapter proposes the concept of aquatic justice, which extends environmental justice principles to aquatic ecosystems. Environmental justice traditionally focuses on the equitable distribution of environmental benefits and burdens among human populations [31]. Aquatic justice expands this framework to include non-human entities, recognizing water-based ecosystems as moral subjects deserving of protection and respect as well.

Aquatic justice is grounded in three major principles, which include:

5.1 Intrinsic value of aquatic ecosystems

Water bodies, from rivers and seas to oceans, are not merely resources but dynamic systems with inherent worth. This aligns with ecocentric ethics and acknowledges the interdependence of aquatic life [27].

5.2 Intergenerational equity

The persistence of microplastics imposes burdens on future generations, which comprise both human and non-human life. Aquatic justice demands policies that prioritize long-term ecological health over short-term economic gains [32].

5.3 Restorative responsibility

Humans must actively and deliberately remediate the harm caused by microplastics through clean-up efforts, innovation in biodegradable materials, and systemic changes in production and consumption. [33].

Aquatic justice, therefore, challenges the anthropocentric bias in environmental policy and calls for a paradigm shift that prioritizes the health of aquatic ecosystems as a moral imperative for the promotion of aquatic life. Key dimensions of environmental decision-making are outlined in **Table 2**.

It is important to state here that, although microplastic pollution is a global issue, its impacts are distributed unevenly across populations. The framework of environmental justice emphasizes not only the equitable distribution of environmental benefits and burdens but also the inclusion of historically marginalized voices in environmental governance. Recognizing the disproportionate vulnerability of certain groups, such as indigenous communities, developing nations,

Principle	Ethical basis	Application example	Reference
Intrinsic value	Ecocentrism	Recognizing coral reefs as valuable beyond tourism.	[27]
Intergenerational equity	Environmental ethics	Regulating microplastic waste to protect future species.	[32]
Restorative responsibility	Kantian/ Utilitarian	Mandating plastic producers to fund clean-up efforts.	[33]

Table 2.
Principles of aquatic justice and their applications.

and low-income coastal populations, is critical to a more inclusive ethical response to microplastic contamination.

Indigenous communities, for example, are disproportionately affected by microplastic pollution due to their close dependence on aquatic ecosystems for cultural, nutritional, and spiritual practices. The contamination of rivers and coastal waters with microplastics threatens traditional food systems, erodes cultural practices, and exacerbates ecological degradation in indigenous territories. These communities often lack formal political representation in environmental decision-making processes, which further compounds their exposure to environmental harm and structural marginalization [34, 35].

In developing countries, the situation is aggravated by weak regulatory systems, inadequate waste management infrastructure, and the outsourcing of plastic waste from developed countries. Many of these nations, particularly in Asia and Africa, serve as dumping sites for global plastic waste, increasing their exposure to microplastic pollution and widening existing inequalities [36, 37]. The absence of recycling infrastructure and poor industrial oversight further intensifies local contamination.

Furthermore, low-income urban communities, especially those in coastal areas, face greater health risks from microplastics due to reliance on contaminated seafood and unfiltered water sources. Exposure to microplastic-related chemicals has been linked to endocrine disruption, inflammation, and oxidative stress, which may be more severe among populations with limited access to healthcare and environmental protections [38, 39].

6. Practical implications and ethical action

Mitigating the microplastic crisis, no doubt, requires integrating ethical principles with evidence-based solutions. Global plastic production exceeds 400 million tonnes annually, with 8–10 million tonnes entering oceans yearly, which necessitates urgent action [12]. Addressing the microplastic crisis, therefore, requires integrating ethical principles into practical solutions, not just drafting theoretical policies that are not pragmatic. For example, policy measures, such as the EU's 2021 single-use plastic ban (reducing plastic bag use by 50%), show progress, but global coordination remains limited [39]. A comparative policy analysis also reveals disparities; for instance, Xanthos and Walker [39] reported a 70% reduction in cosmetic microplastics post-Canada's microbead ban, while weaker enforcement in Southeast Asia yielded minimal impact [39]. Moreover, technological innovations, like nanofiltration systems that removed 99% of microplastics from wastewater, proved promising but require scaling [40]. Ethical action involves corporate accountability (e.g., biodegradable packaging mandates reducing plastic waste by 30% in pilot studies) and consumer responsibility [12]. Public awareness campaigns, addressing 80% unawareness of microplastic sources, can drive behavioral change, with surveys showing a 60% increase in recycling post-education [41].

Philosophically, virtue ethics offers a complementary approach, emphasizing the cultivation of environmental virtues such as humility, prudence, and care [35]. It is believed that by fostering these virtues, individuals and societies can shift toward sustainable practices that mitigate the microplastic crisis.

Therefore, to address the escalating microplastic crisis, we propose that policy actors, educators, and non-governmental organizations (NGOs) must ground their interventions in ethical reasoning as well as empirical data. Doing so can strengthen both the ecological and social dimensions of sustainability. The following recommendations put forward outline practical steps that balance environmental integrity with social justice.

6.1 Strengthen regulation of plastics and waste systems

Governments should restrict single-use plastics and improve waste management infrastructure, while also ensuring that the benefits and burdens of such measures are fairly distributed. Evidence from the European Union's 2021 plastic restrictions, which cut plastic bag use by half, demonstrates the effectiveness of such policies [39]. Equally important is the equitable enforcement of regulations in low- and middle-income countries, where the importation of plastic waste compounds existing environmental pressures [36].

6.2 Adopt aquatic justice as a policy principle

Policymakers should explicitly recognize aquatic justice as an approach that values the inherent worth of aquatic ecosystems while promoting fairness for communities dependent on them. This perspective is consistent with ecocentric ethics, which protects marine biodiversity while also addressing the unequal effects of pollution on vulnerable populations [7, 27].

6.3 Integrate environmental ethics into education

Schools and civic education initiatives should include modules or curricula on environmental ethics and intersectionality to make clear the links between everyday consumption and ecological harm. For instance, plastic pollution is estimated to kill 1.5 million marine animals each year [15]. Teaching methods that emphasize biocentric and ecocentric perspectives can help cultivate a sense of moral responsibility for aquatic life.

6.4 Promote intergenerational sustainability learning

Public education should encourage reflection on the long-term consequences of plastic pollution, which can persist in aquatic environments for centuries. By framing sustainability as an intergenerational obligation, such initiatives can help foster attitudes of stewardship across different age groups [13, 32].

6.5 Design community-specific outreach programs

Awareness campaigns are most effective when they draw on local contexts. For example, in some coastal waters, concentrations range from 10 to 100 microplastic particles per liter – a fact that can help communities recognize the immediacy of the threat [42]. Programs rooted in local realities can also equip communities with strategies for reducing plastic use and demanding policy change.

6.6 Elevate marginalized perspectives

Communities that are most at risk of microplastic exposure, such as indigenous groups and those reliant on local fisheries, should be at the center of decision-making. Their exposure is often 30% higher than average due to dependence on contaminated resources [34]. Incorporating indigenous knowledge and ensuring active representation in governance can enrich both the fairness and the effectiveness of policies [34].

6.7 Work with grassroots and coastal organizations

Collaborating with fishing groups and local associations can generate practical solutions, such as alternative fishing gear materials or recycling programs for nets. Since discarded nets account for up to one-fifth of marine plastic debris, such efforts can make a measurable difference [23].

6.8 Advance biocentric and ecocentric policy reform

Moving from an anthropocentric mind-set toward biocentric and ecocentric frameworks is essential for long-term solutions. This includes support for biodegradable alternatives, which, in pilot studies, have cut plastic waste by almost a third [12]. In prioritizing ecological health over short-term economic benefits, such policies also serve the broader aim of justice for human and non-human life alike.

Taken together, these recommendations highlight the importance of combining ethical commitments with scientific knowledge. Coordinated action across education, policy, and community spheres is obviously needed to confront the microplastic crisis and to secure a future in which both human communities and aquatic ecosystems can flourish.

Table 3 below, therefore, outlines a multilevel approach to addressing microplastic pollution, showing how different actors, such as governments, industries, communities, scientists, and philosophers, can contribute to solutions for microplastic pollution. The table illustrates that solving microplastic pollution requires coordinated efforts across sectors, combining laws, technology, education, industry reforms, and ethical consciousness.

7. Conclusion

From the foregoing, it is seen that the microplastic crisis is not merely an environmental issue but a profound ethical challenge that demands a re-evaluation of human responsibility toward aquatic ecosystems. It is opined that by applying philosophical frameworks such as anthropocentrism, biocentrism, and ecocentrism, we can gain a nuanced understanding of the moral dimensions of this crisis. The concept of aquatic justice provides a novel framework for addressing the intrinsic value of water-based ecosystems, intergenerational equity, and restorative responsibility. Hence, as humanity therefore continues to grapple with the consequences of its actions, environmental ethics offers a guiding light, a pathway for fostering a just and sustainable eco-friendly relationship with the aquatic world. Through collective action, therefore, informed by philosophical inquiry, we can

Strategy	Level of action	Example	Quantitative impact	Reference
Policy intervention	Government	Ban on single-use plastics.	50% reduction in plastic bag use (EU, 2021).	[39]
Corporate accountability/corporate social responsibility	Industry	Biodegradable packaging mandates should be upheld and encouraged.	30% reduction in plastic waste (pilot studies).	[12]
Public awareness and education	Community/ society	Campaigns about microplastic impacts are mandatory to be carried out.	60% increase in recycling rates post-campaign.	[41]
Technological innovation	Science/ engineering	Development of filtration systems in laundry becomes a sine-qua-non.	99% microplastic removal efficiency.	[40]
Ethical reorientation	Philosophical/ cultural	Promotion of eco-virtues, like prudence and care, is necessary for the safety of ecosystems.	20% shift to sustainable behaviors (surveys).	[43]

Table 3.
Strategies for mitigating microplastic pollution.

mitigate the microplastic crisis and uphold our moral obligations to the planet and its inhabitants for a more sustainable life.

AI usage declaration/acknowledgment

The authors of this manuscript (i.e., Oviemuno Wilfred Egara, Dr. Peter M. Etaware, and Bridget Gwa Oviemuno) declare that AI tools were slightly employed in putting this work together. QuillBot was seldom used as a paraphraser in order to avoid plagiarism and enhance content originality, while Grammarly was used for proper sentence construction to enhance the presentation of the work for readability purposes. ChatGPT was used for fact-checking and comparison. The content of the manuscript is based on the personal effort of all authors concerned.

Thank you.

Author details

Egara Wilfred Oviemuno[1]*, Etaware Peter Mudiaga[2] and Oviemuno Bridget Gwa[3]

1 Directorate of General Studies, Southern Delta University, Ozoro, Nigeria

2 Plant Science and Biotechnology Unit, Department of Biological Sciences, Faculty of Science, Southern Delta University, Ozoro, Nigeria

3 Department of Business Management, Faculty of Social Sciences, Moses Adasu University, Makurdi, Nigeria

*Address all correspondence to: egaraovie@gmail.com

References

[1] Cole M, Lindeque P, Halsband C, Galloway TS. Microplastics as contaminants in the marine environment: A review. Marine Pollution Bulletin. 2011;**62**(12): 2588–2597. DOI: 10.1016/j.marpolbul.2011.09.025

[2] Van Sebille E, Wilcox C, Lebreton L, Maximenko N, Hardesty BD, van Franeker JA, Eriksen M, Siegel D, Galgani F, Law KL. A Global inventory of small floating plastic debris. Environmental Research Letters. 2015;**10**(12):124006. DOI: 10.1088/1748-9326/10/12/124006

[3] Rochman CM, Browne MA, Underwood AJ, van Franeker JA, Thompson RC, Amaral-Zettler LA. The ecological impacts of marine debris: Unraveling the demonstrated evidence from what is perceived. Ecology. 2016;**97**(2):302–312. DOI: 10.1890/14-2070.1

[4] Barboza LGA, Vethaak AD, Lavorante BR, Lundebye AK, Guilhermino L. Marine microplastic debris: An emerging issue for food security, food safety and human health. Marine Pollution Bulletin. 2018;**133**:336–348

[5] Lusher AL, Hollman PCH, Mendoza-Hill JJ (2017). Microplastics in fisheries and aquaculture: Status of knowledge on their occurrence and implications for aquatic organisms and food safety. FAO Fisheries and Aquaculture Technical Paper. **615**. Rome: FAO

[6] Schlosberg D. Defining Environmental Justice: Theories, Movements, and Nature. Oxford: Oxford University Press; 2020

[7] H III. R. Environmental ethics: Values in and duties to the natural world. In: Bormann FH, Kellert SR, editors. The Broken Circle: Ecology, Economics, Ethics. New Haven: Yale University Press 1991;73–96

[8] GESAMP. Sources, Fate and Effects of Microplastics in the Marine Environment: Part 2 of a Global Assessment. Joint Group of Experts on the Scientific Aspects of Marine Environmental Protection. A report to inform the Second United Nations Environment Assembly GESAMP Working Group 40, 2nd phase. London: IMO. 2021

[9] Obbard RW, Sadri S, Wong YQ, Khitun AA, Baker I, Thompson RC. Global Warming Releases Microplastic Legacy Frozen in Arctic Sea Ice. Earth's Future. 2014;**2**(6):315–320. DOI: 10.1002/2014EF000240

[10] Napper IE, Thompson RC. Release of synthetic microplastic plastic fibres from domestic washing machines: Effects of fabric type and washing conditions. Marine Pollution Bulletin. 2016;**112**(1-2):39–45. DOI: 10.1016/j.marpolbul.2016.09.025

[11] Lebreton LCM, van der Zwet J, Damsteeg JW, Slat B, Andrady A, Reisser J. River plastic emissions to the world's oceans. Nature Communications. 2017;**8**:15611. DOI: 10.1038/ncomms15611

[12] Jambeck JR, Geyer R, Wilcox C, Siegler TR, Perryman M, Andrady A, et al. Plastic waste inputs from land into

the ocean. Science. 2015;**347**(6223): 768–771. DOI: 10.1126/science.1260352

[13] Barnes DKA, Galgani F, Thompson RC, Barlaz M. Accumulation and Fragmentation of Plastic Debris in Global Environments. Philosophical Transactions of the Royal Society B: Biological Sciences. 2009;**364** (1526):1985–1998. DOI: 10.1098/rstb.2008.0205

[14] Wright SL, Thompson RC, Galloway TS. The physical impacts of microplastics on marine organisms: A review. Environmental Pollution. 2013;**178**:483–492. DOI: 10.1016/j.envpol.2013.02.031

[15] Gall SC, Thompson RC. The Impact of Debris on Marine Life. Marine Pollution Bulletin. 2015;**92**(1-2): 170–179. DOI: 10.1016/j.marpolbul.2014.12.041

[16] Kosuth M, Mason SA, Wattenberg EV. Anthropogenic contamination of tap water, beer, and sea salt. PLoS One. 2018;**13**(4): e0194970. DOI: 10.1371/journal.pone.0194970

[17] Shim WJ, Hong SH, Eo SE. Identification methods in microplastic analysis: A review. Analytical Methods. 2017;**9**(9):1384–1391. DOI: 10.1039/C6AY02558G

[18] Ziajahromi S, Neale PA, Leusch FDL. Wastewater treatment plant effluent as a source of microplastics: Review of the fate, chemical interactions and potential risks to Aquatic organisms. Water Science and Technology. 2016;**74** (10):2253–2269. DOI: 10.2166/wst.2016.414

[19] Hidalgo-Ruz V, Gutow L, Thompson RC, Thiel M. Microplastics in the marine environment: A review of the methods used for identification and quantification. Environmental Science & Technology. 2012;**46**(6):3060–3075. DOI: 10.1021/es2031505

[20] GESAMP. Guidelines for the monitoring and assessment of plastic litter and microplastics in the ocean. Joint Group of Experts on the Scientific Aspects of Marine Environmental Protection. Reports and Studies 99. London: IMO. 2021

[21] Karlsson TM, Arneborg L, Broström G, Almroth BC, Gipperth L, Hassellöv M. The unaccountability case of plastic pellet pollution. Marine Pollution Bulletin. 2018;**129**(1):52–60. DOI: 10.1016/j.marpolbul.2018.01.041

[22] Wagner M, Scherer C, Alvarez-Muñoz D, Brennholt N, Bourrain X, Buchinger S, et al. Microplastics in freshwater ecosystems: What we know and what we need to know. Environmental Sciences Europe. 2014;**26**(1):12. DOI: 10.1186/s12302-014-0012-7

[23] FAO. The state of world fisheries and aquaculture 2020: sustainability in action. Rome: FAO; 2020. DOI: 10.4060/ca9229en

[24] UNEP. From pollution to solution: a global assessment of marine litter and plastic pollution. Nairobi: United Nations Environment Programme; 2021

[25] Beaumont NJ, Aanesen M, Austen MC, Börger T, Clark JR, Cole M, Hooper T, Lindeque PK, Pascoe C, Wyles KJ. Global Ecological, Social and Economic Impacts of Marine Plastic. Marine Pollution Bulletin.

2019;**142**:189–195. DOI: 10.1016/j.marpolbul.2019.03.022

[26] Taylor PW. Respect for Nature: A Theory of Environmental Ethics. USA: Princeton University Press; 1986

[27] Leopold A. A Sand County Almanac and Sketches Here and There. USA: Oxford University Press; 1949

[28] Kant I. Groundwork of the Metaphysics of Morals. Gregor M, Trans. United Kingdom: Cambridge University Press; 1998. (Original work published 1785)

[29] Jonas H. The Imperative of Responsibility: In Search of an Ethics for the Technological Age. USA: University of Chicago Press; 1984

[30] Nixon R. Slow Violence and the Environmentalism of the Poor. USA: Harvard University Press; 2011

[31] Bullard RD. Dumping in Dixie: Race, Class, and Environmental Quality. New York: Routledge; 2018. DOI: 10.4324/9780429495274

[32] Weiss EB. In Fairness to Future Generations: International Law, Common Patrimony, and Intergenerational Equity. USA: Transnational Publishers; 1989

[33] Jamieson D. When Utilitarians should be Virtue Theorists. Utilitas. 2007;**19**(2):160–183. DOI: 10.1017/S0953820807002452

[34] Kameri-Mbote P, Kaguru M. The Role of Indigenous and traditional knowledge in combating plastic pollution. Strathmore Law Journal. 2025;75–90. DOI: 10.52907/slj.v8i1.534

[35] K.p W. Indigenous climate change studies: Indigenizing futures, decolonizing the Anthropocene. English Language Notes. 2017;**55**(1-2):153–162

[36] Brooks AL, Wang S, Jambeck JR. The Chinese import ban and its impact on global plastic waste trade. Science Advances. 2018;**4**(6):1–7

[37] Da Costa JP, Mouneyrac C, Costa M, Duarte AC, Rocha-Santos T. The role of legislation, regulatory initiatives and guidelines on the control of microplastics: The case of the European Union. Microplastics and Nanoplastics. 2020;**1**(1):1–16

[38] J.c P. Microplastics in wastewater: State of the knowledge on sources, fate and solutions. Marine Pollution Bulletin. 2018;**129**(1):262–265

[39] Xanthos D, Walker TR. International policies to reduce plastic marine pollution from single-use plastics (Plastic Bags and Microbeads): A review. Marine Pollution Bulletin. 2017;**118**(1-2):17–26. DOI: 10.1016/j.marpolbul.2017.02.048

[40] Enfrin M, Dumée LF, Lee J. Nano/microplastics in water and wastewater treatment processes – origin, impact and potential solutions. Water Resources. 2019;**161**:621–638. DOI: 10.1016/j.watres.2019.06.025

[41] Pahl S, Wyles KJ, Thompson RC. Channelling passion for the ocean towards plastic pollution. Nature Human Behaviour. 2017;**1**(10):697–699. DOI: 10.1038/s41562-017-0204-4

[42] Li WC, Tse HF, Fok L. Plastic waste in the marine environment: A review of sources, occurrence and effects. Science of the Total

Environment. 2016;**566-567**:333–349. DOI: 10.1016/j.scitotenv.2016.05.084

[43] Cafaro P. Thoreau, Leopold, and Carson: Toward an Environmental Virtue Ethics. Environmental Ethics. 2001;**23**(1):3–17. DOI: 10.5840/enviroethics200123135